Percy Ludgate (1883-1922):
Ireland's First Computer Designer

First Edition

Edited by

Brian Coghlan and Brian Randell

with the assistance of

Paul Hockie, Trish Gonzalez, David McQuillan & Reddy O'Regan

Published by The John Gabriel Byrne Computer Science Collection.
First published in Dublin, Republic of Ireland, 2022

ISBN: 978-1-911566-29-8

Typeset in Times New Roman

Printed and bound by IngramSpark

*"I PURPOSE to give in this paper a short account of the result of about six years' work,
undertaken by me with the object of designing machinery capable of performing calculations,
however intricate or laborious, without the immediate guidance of the human intellect."*
from Percy Ludgate's 1909 paper.

Table of Contents

Introduction

The 16th day of October, 2022, will be the 100th anniversary of the death of Percy Edwin Ludgate at his former home at 30 Dargle Road, Drumcondra, where in his evening hours he designed the World's second computer, what he called his *Analytical Machine*. His design was published in 1909, sixty-six years after Ada Lovelace published the first design for a computer, Charles Babbage's famous *Analytical Engine*.

In 1970 Brian Randell stumbled across Ludgate shortly after arriving at Newcastle University, while he was preparing his professorial Inaugural Lecture and was looking for interesting things to say about Ada Lovelace (1815-52) and Charles Babbage (1791-1871).

Babbage was a brilliant mathematician, philosopher, inventor and mechanical engineer, who originated the concept of a programmable digital computer. He spent many years and a small fortune designing and trying (unsuccessfully) to construct machines for calculating and printing mathematical tables, and from 1834 his Analytical Engine – a giant mechanical computer controlled by punched card programs, the forerunner of modern electronic computers. This was a monstrous machine, of which only a small fragment was ever built.

Ada was the daughter of Lord Byron, who was absent from shortly after her birth. She was brought up by Lady Byron, who promoted her interest in mathematics, aiming to avoid Ada developing her father's notorious character. At the age of 18, when she was taken to a Babbage soirée, she became fascinated by his work on calculating machines. She has become famous for her translation and extensive annotation of an Italian description of the Analytical Engine, going beyond Babbage in documenting the Engine's programming and conveying its potential. She died tragically early of uterine cancer aged 36, but is now a veritable icon.

Randell's search for something new to say about Ada Lovelace led him to the 1926 London Science Museum catalogue of *Calculating Machines and Instruments*, where he noticed it mentioned that a Percy Ludgate, an unfamiliar name, had designed a Difference Engine. Exploring further, he was led to the splendid *Handbook of the 1914 Napier Tercentenary Exhibition*, in which he found Percy Ludgate had a Chapter: "Automatic Calculating Machines" which ended with this startling claim: "I have myself designed an analytical machine", and a reference to Ludgate's 1909 paper in the *Scientific Proceedings of the Royal Dublin Society*. The paper's opening sentence begins: "I purpose to give in this paper a short account of the result of about six years' work".

By 1909, calculating machines – first invented in the 17th century – were by now widely available to (rich) businesses and scientists. Few performed automatic multiplication. Adding machines (invented in the late 19th century) were coming into general use in well-equipped businesses and commerce. Punched card machines, invented for tabulating the 1890 U.S. census, had been used for many countries' national censuses, and were starting to be used for other large data processing tasks by a few large organisations. But the notion of an automatic, i.e. programmable, calculator had been largely forgotten.

Ludgate's paper credibly claims little prior knowledge of Babbage – the three main components (store, arithmetic unit, sequence controller) of his machine all show evidence of considerable ingenuity and originality. Numbers were to be represented by sliding rods in rotatable storage cylinders. Arithmetic was to be performed using what were later called *Irish Logarithms*. Program sequencing was to be controlled from keyboards or perforated papers. But he never constructed his machine, and thus far only one drawing of it has been found.

Randell's investigation tracked down just one colleague with recollections of Ludgate as an accountant, and just one relative of Percy Ludgate, his niece Violet. The latter was actually found by the Secretary of the Royal Dublin Society, who called all the Ludgates in the Dublin telephone directories! Violet's family had lived near Percy Ludgate – she was 19 when he died – her memories were the main source of information, and she provided the only known photograph of Ludgate.

In 1991 an academic prize was established in Ludgate's honour. Then in 2016 an investigation was begun by Brian Coghlan under the auspices of The John Gabriel Byrne Computer Science Collection, which now holds copies (many original) of all the known literature and records relating to Ludgate. Its cataloguing had prompted the investigation, beginning with assembly of a team of experts in the fields thought necessary: computer science, history of computing, mathematics, genealogy and law. The aims included whether it was possible to discover any of Ludgate's drawings and documents, descendents of his extended family, relevant documents, photos or memories. The results of this and previous investigations are described below, along with Ludgate's original papers.

It was to be another forty years after Ludgate's paper before the world's first practical electronic programmable (i.e. general purpose) computer became operational, the EDSAC at Cambridge, in May 1949. There are no known links back from the first electronic computers to the prior work of Ludgate, or that of the few other pioneers known to have designed (mechanical or electromechanical) computers.

The material below is structured to be easily digestible by the general readership, human interest first, technicalities next, gradually increasing in depth. This results in a variant of reverse chronological order of material, culminating in Percy Ludgate's original paper.

Our sincere thanks to the members of the investigation team, and to the numerous individuals and organisations who have helped our very extensive search for further information about Ludgate and his work. Finally we extend our thanks for the support of the School of Computer Science and Statistics, Trinity College Dublin, for this work and for *The John Gabriel Byrne Computer Science Collection*.

Brian Coghlan and Brian Randell, 2022

To Dig Deeper, browse to: https://www.scss.tcd.ie/SCSSTreasuresCatalog/ludgate

The Percy E. Ludgate Prize in Computer Science

This academic prize was instituted in 1991 by Professor John Gabriel Byrne of the Department of Computer Science at Trinity College Dublin, to be awarded annually to the student who submits the best project in the senior year of the Moderatorship in Computer Science. The annual Court of Examiners for Computer Science nominates the recipient each year on the basis of the student's Final Year Project mark, the major component of which is a substantial project report describing individual work done by the student under the supervision of a member of staff. The project is one component of the overall final year curriculum. There is no ceremony for the awarding of the prize and, as far as is known, prizes are not mentioned at commencement (graduation) ceremonies.

The Prize is a cash prize (€127) transferred to the recipient's bank account. The source of funds each year is a benefaction from a specific Fund (number BO0333), the original source of which is not yet known. The School pays the recipient each year directly to their bank account from that Fund. In recent years at least, the recipients are listed by Trinity College Dublin in its *Calendar*. The prizewinners thus far are listed below.

Year	Surname	Forename	Year	Surname	Forename
1990/1991	Collins	Steven	2005/2006	McCabe	Eoghan
1991/1992	Reynolds	Hugh	2006/2007	Kelly	Conor
1992/1993	Gallagher	James	2007/2008	Doyle	Michael
1993/1994	Cassidy	Garrett	2008/2009	Purcell	Stephen
1994/1995	Courtney	Anthony	2009/2010	Lynch	Aidan
	Peirce	Michael		Clear	Michael
1995/1996	Murray	Damien	2010/2011	Dolan	Stephen
1996/1997	Boyle	Peter	2011/2012	Browne	Don
	Brennan	James	2012/2013	Hughes	Macdara
	Kaatz	Rudiger	2013/2014	Rogers	Stephen
1997/1998	Feeney	Kevin	2014/2015	Donnelly	Matthew
1998/1999	McConnell	Walter	2015/2016	McNamee	Eamonn
1999/2000	Dunne	Kevin	2016/2017	Leacy	David
2000/2001	Hamill	John	2017/2018	Corkery	Gavin
2001/2002	Clarke	Niall	2018/2019	Mongan	Aidan
2002/2003	Caulfield	Darren	2019/2020	Carbeck	John
	Lee	Richard	2020/2021	Suliman	Mohamed
2003/2004	De Vries	Edsko			
2004/2005	Woulfe	Muiris			

Percy Ludgate (1883–1922), Ireland's first computer designer

BRIAN COGHLAN*
School of Computer Science and Statistics, Trinity College Dublin, Ireland.

BRIAN RANDELL
School of Computing, Newcastle University, U.K.

PAUL HOCKIE
London, U.K.

TRISH GONZALEZ
Florida, U.S.A.

DAVID MCQUILLAN
Wokingham, U.K.

REDDY O'REGAN
Skibbereen, Co.Cork, Ireland.

[Accepted 21 May 2021. Published 9 September 2021.]

Abstract

A greatly expanded treatment is presented of the history of the family, life and work of Percy Ludgate, nearly 50 years after the 1971 and 1982 papers by Brian Randell revealed his work on a mechanical computer, and almost 100 years after Ludgate's death. The new material that has recently been obtained about this successor of Charles Babbage includes two very significant discoveries. The first is of a hitherto unknown contemporary published description of Ludgate's Analytical Machine, incorporating the only surviving drawing of it yet found; the second is of American descendants of Ludgate's niece, who have been allowed to erect a commemorative headstone on his previously unmarked grave.

*Author's email: coghlan@cs.tcd.ie.
ORCID iD: https://orcid.org/0000-0002-1725-1362
doi: https://doi.org/10.3318/PRIAC.2021.121.09

Proceedings of the Royal Irish Academy Vol. 121C, 303–332 © 2021 The Author(s).

Brian Coghlan et al.

Introduction

The story of how in 1834, over a hundred years before the first electronic computers, the brilliant but eccentric English scientist, Charles Babbage, started to design a huge sequence-controlled mechanical digital computer,[1] his 'Analytical Engine',[2] to improve on the human 'computers' then employed,[3] is reasonably well known. However, less is known of Percy Edwin Ludgate, who was the second person to design a sequence-controlled mechanical digital computer in the pre-electronic era.

Remarkably, Ludgate, who was born in Skibbereen in 1883 and employed in Dublin until his death in 1922, was just a clerk to a corn merchant and subsequently a qualified accountant. He worked on his design in his spare time between 1903 and 1909, and was not aware of Babbage's work until his initial design was completed. His 'Analytical Machine',[4] as he called it, was quite unlike Babbage's in that it was largely based on multiplication using novel mechanical ideas, while Babbage's was very different mechanically, and was, more conventionally, based on addition. Brian Randell's 1982 paper states that:

> although Ludgate had, at least during the later stages of his work, known of Babbage's machine, much of his work was clearly entirely original – and indeed with respect to program control, a distinct advance on Babbage's ideas. In fact, all three main components of Ludgate's Analytical Machine – the store, the arithmetic unit, and the sequencing mechanism – show evidence of considerable ingenuity and originality.[5]

Both machines embraced the core concept of being able to be automatically sequenced through a set of operations,[6] using perforated paper in Ludgate's case, Jacquard-style strung-together punched cards in Babbage's, and in particular both machines supported what is now known to be the critical concept of being able to change the sequencing mechanism's behaviour based on the results of

---

[1] A 'sequence-controlled digital computer' is known as a 'program-controlled digital computer' or 'programmable digital computer', typically abbreviated to 'digital computer' in modern terminology.

[2] Ada Lovelace, 'Sketch of the analytical engine invented by Charles Babbage, esq. by L.F. Menabrea, of Turin, officer of the military engineers, with notes by the translator', in Richard Taylor (ed.), *Scientific memoirs* (London, 1843), vol. 3, 666–731. Available at: https://www.scss.tcd.ie/SCSSTreasuresCatalog/literature/TCD-SCSS-V.20121208.870/TCD-SCSS-V.20121208.870.pdf.

[3] A term used since the early seventeenth century for a person performing mathematical calculations.

[4] Percy E. Ludgate, 'On a proposed analytical machine', *Scientific Proceedings of the Royal Dublin Society* 12:9 (Dublin, 28 April 1909), 77–91. Available at: https://www.scss.tcd.ie/SCSSTreasuresCatalog/literature/TCD-SCSS-V.20121208.873/TCD-SCSS-V.20121208.873.pdf. Also reproduced in Randell's 1971 paper, see n. 8 for details.

[5] Brian Randell, 'From analytical engine to electronic digital computer: the contributions of Ludgate, Torres, and Bush', *IEEE Annals of the History of Computing* 4:4 (1982), 327–41.

[6] Otherwise referred to here as 'sequence-controlled', i.e., 'program-controlled' or 'programmable' (see n. 1), a core concept for modern computers.

prior calculations.[7] Thus both machines would be capable of doing everything a modern computer of their scale and class could do, although very slowly. If they had not incorporated these concepts, both the designs and their designers would be much less significant with respect to the history of computing.

Ludgate died in 1922, his work already having faded into obscurity. Almost nothing further was heard of him until Randell's 1971 and 1982 papers,[8] and Riches' 1973 project.[9] Since then Ludgate's work has often been described in general accounts of the history of computing, and there have been many summaries of Randell's papers, in, for example, the *Dictionary of Irish Biography*,[10] although nothing new was published until Jim Byrne's 2011 paper provided some extra detail on Ludgate's parents.[11] Meanwhile, in 1991, the Department of Computer Science in Trinity College Dublin instigated an annual prize in his memory,[12] the only official recognition worldwide until recently.

An investigation was initiated in 2016, under the aegis of the John Gabriel Byrne Computer Science Collection,[13] aimed at fully exploring and documenting Ludgate's life and work.[14] To that end, information relating to Ludgate's own life and early career, his close family members and their children, and his ancestors back to the early eighteenth century, domestic situation, legal issues, and 1909 paper submission, have all been examined, and one important discovery has been made concerning his family. Moreover, joint outreach activities

---

[7] 'Conditional-branching' in modern terminology, a critical concept for modern computers.

[8] Brian Randell, 'Ludgate's analytical machine of 1909', *The Computer Journal* 14:3 (1971), 317–26. For Randell's 1982 paper see n. 5 above.

[9] D. Riches, 'An analysis of Ludgate's machine leading to the design of a digital logarithmic multiplier' (unpublished undergraduate project report), Department of Electrical and Electronic Engineering, University College (Swansea), June 1973. Available at n. 14 below.

[10] Brian Randell, 'Ludgate, Percy', *Dictionary of Irish Biography* (9 vols, Cambridge, 2009). Available at: https://www.dib.ie/biography/ludgate-percy-a4912.

[11] Jim Byrne, 'Percy E. Ludgate (1883–1922): Skibbereen-born computer pioneer', *Skibbereen and District Historical Society Journal* 7 (2011), 72–80.

[12] The Percy E. Ludgate Prize in Computer Science is awarded annually to the student who submits the best project in the senior year of the Moderatorship in Computer Science at Trinity College Dublin (TCD).

[13] The John Gabriel Byrne Computer Science Collection at TCD, some 3000 items, holds a wealth of literature (including original offprints of Lovelace's 1843 monograph and of Ludgate's 1909 paper), plus a wide range of computers, calculating machines, slide rules and computer software, and is recorded in the Schedules to the Statutes as being among the treasures of the College. Copies (in most cases original) of all the known literature and records relating to Ludgate are now held within this collection. The catalogue is available at: https://www.scss.tcd.ie/SCSSTreasuresCatalog/.

[14] Ludgate folder at TCD. Available at: https://www.scss.tcd.ie/SCSSTreasuresCatalog/miscellany/TCD-SCSS-X.20121208.002/. For a more mobile-friendly interface to selected Ludgate folder contents, see: https://www.scss.tcd.ie/SCSSTreasuresCatalog/ludgate/. Copies of most of the documents cited in this paper are available at this folder.

with the Ludgate Hub,[15] both in newspapers[16] and at the West Cork History Festival,[17] led to a second very significant discovery: that of the first known contemporary drawing of part of Ludgate's machine. As a result, it has been possible to gain a new depth of understanding of key mechanisms within his machine,[18] and to attempt to codify this,[19] even though his extensive set of drawings (which he stated, in both his 1909 and 1914 papers, that he had made) has never been found. It is now possible to provide a greatly expanded summary of what is known of the family, life and work of Ludgate, giving a whole range of new material (all older material from Randell's 1971 paper is explicitly identified to clearly distinguish it from recent research).

In his 1971 investigations Randell established contact with one of Ludgate's accountancy colleagues, who provided a limited amount of information. But the great majority of what was discovered then of Ludgate's family and life was obtained from just one person, his niece Violet Ludgate, who Randell stated 'made extensive efforts herself to trace further possible sources of information'. Nonetheless the family details uncovered related just to Percy Ludgate's

PL. I—The eighteenth-century farmhouse (with 4ft-thick walls) of Marble Hill Farm, Kilshannig, Co. Cork (Image courtesy of the John Gabriel Byrne Computer Science Collection).

---

[15] An e-commerce hub opened in 2016 in Ludgate's birthplace Skibbereen, see: Noel Baker, 'Potential €37m impact of digital life on Skibbereen', *Irish Examiner*, 5 November 2015, 7.

[16] David Forsythe, 'Could WW1 have ended Ludgate's computer dreams?', *The Southern Star*, 27 July 2019, 11; Kevin O'Neill, 'West Cork's own computer pioneer', *Cork Examiner*, 6 August 2019, 2.

[17] Brian Coghlan, 'An exploration of the life of Percy Ludgate', *West Cork History Festival* (10 August 2019).

[18] Brian Coghlan, Brian Randell, Paul Hockie, Trish Gonzalez, David McQuillan, Reddy O'Regan, 'Investigating the work and life of Percy Ludgate', *IEEE Annals of the History of Computing* 43:1 (2021), 19–37.

[19] Brian Coghlan, 'Percy Ludgate's analytical machine'. Available at n.14 above.

Percy Ludgate (1883–1922), Ireland's first computer designer

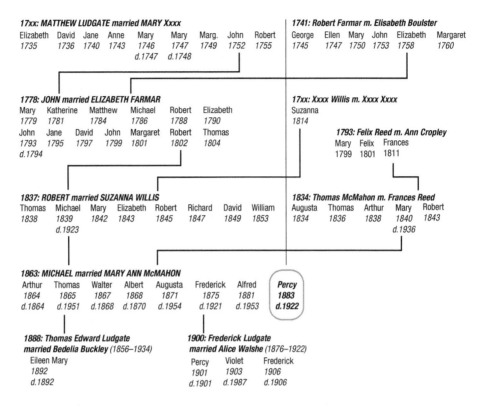

Fig. 1—Summary Family Tree of Michael Ludgate and Mary Ludgate née McMahon (Image courtesy of the John Gabriel Byrne Computer Science Collection. © Brian Coghlan 2020).

parents and siblings but even missed his sister. Percy Ludgate's paternal ancestors have now been traced to the early 1700s, when Matthew and Mary Ludgate (Percy's paternal great-great-grandparents) are believed to have leased Marble Hill Farm, in the townland of Scarragh, Kilshannig, three miles west of Mallow, Co. Cork, see Plate I.[20]

Matthew Ludgate appears in the 1766 religious census for Kilshannig.[21] The baptisms, burials and marriages of his descendants (see Fig. 1) are recorded in Mallow Church of Ireland records, which began in 1731. A very extensive family tree of about 100 descendants of Matthew and Mary can be constructed from these records.[22] Of these, their son John and grandson Robert (paternal

---

[20] Brian Coghlan, Personal communication with owner of Marble Hill Farm (August 2019).
[21] National Archives of Ireland (NAI), Return no. 1123 (transcript), MS 5036(a). Religious Census 1766 for Kilshannig, Cloyne.
[22] 'Percy E. Ludgate Prize in Computer Science' (see n. 14). This is the primary biographical and genealogical reference document. For genealogy, refer specifically to sections: (a) *Family tree of Percy Ludgate's wider Ludgate relations*; (b) *Ancestors of Percy Edwin Ludgate*; (c) *Ancestors of Barbara Hopkins*; (d) *Ancestors of Eileen Mary Ludgate*.

Brian Coghlan et al.

great-grandfather and grandfather of Percy) established their own farms nearby in the townland of Kilshannig.[23]

Ludgate's maternal ancestors have also been traced, mostly to the eighteenth century, but less comprehensively, again see Figure 1. It is not known where his paternal grandmother Suzanna Willis was born. His maternal grandfather Thomas McMahon was born in County Armagh, and his maternal grandmother Frances Reed was probably born in south-west County Cork.[24]

Via his parent's siblings, and his own siblings, Percy had many close relatives and in-laws. In total on his paternal side he had five uncles, two aunts, and at least seven cousins, while on his maternal side he had three uncles, one maternal aunt, and perhaps cousins. In addition, his sister-in-law Bedelia (Bridget) née Buckley, who was from Douglas, Co. Cork, had eleven siblings, and his other sister-in-law Alice (Alicia) née Walshe from Dublin City had two siblings. In total his close relatives and in-laws included fifteen of his parent's generation, plus at least twenty of his own generation, all born in the nineteenth century, almost all in the Victorian age of empire and accelerating scientific sophistication.

The eighteenth and nineteenth centuries through which these ancestors of Percy Ludgate lived were two very different scientific eras in the UK. The eighteenth century was the last in which science was primarily conducted by financially independent individuals, and when the more deductive scientific method of 'natural philosophy' held sway, whereby derived facts were logically deduced from existing 'knowledge'.[25] From the early nineteenth century a more inductive and disciplined scientific method based on precepts outlined by Francis Bacon progressively dominated, whereby collected evidence could by induction engender falsifiable theories and thereby 'new knowledge' until proven otherwise.[26]

Charles Babbage was closely involved in this change, which was passionately promoted in the first half of the nineteenth century by Babbage, William Whewell, John Herschel and Richard Jones (members of the so-called 'Philosophical Breakfast Club').[27] In 1812, Babbage, Herschel and George Peacock established the Analytical Society to pursue (successfully) the move from Newton's to Leibniz's notation for calculus. In 1820, Babbage, Herschel and others founded the Royal Astronomical Society. From 1822, Babbage began designing his Difference Engine No.1.[28] In 1827, Babbage, with help from an Irish

---

[23] *Griffith Valuation, Parish of Kilshannig.* Available at: http://www.askaboutireland.ie/griffith-valuation/.

[24] 'Percy E. Ludgate Prize in Computer Science' (see n. 22).

[25] Simon Schaffer, 'Scientific discoveries and the end of Natural Philosophy', *Social Studies of Science* 16:3 (1986), 387–420.

[26] William Whewell, *History of the inductive sciences, from the earliest to the present time* (3 vols, London, 1837).

[27] Laura J. Snyder, *The Philosophical Breakfast Club: four remarkable friends who transformed science and changed the world* (New York, 2011).

[28] Babbage was also the first person to design what he called a 'difference engine', a large machine consisting of a number of linked adding mechanisms capable of calculating and printing mathematical tables of polynomial functions.

Percy Ludgate (1883–1922), Ireland's first computer designer

friend,[29] Thomas Colby, published a much-improved table of logarithms.[30] In 1831, William Harcourt, David Brewster, Babbage, Whewell and others founded the British Association for the Advancement of Science (Whewell coined the term 'scientist' at its 1833 meeting). From 1834, Babbage began designing his Analytical Engine.[31] In 1834, Babbage, Jones, Whewell and others founded the Royal Statistical Society. In 1837, Babbage published his controversial *The Ninth Bridgewater Treatise*, employing difference engine concepts in a discussion of how science and religion interrelate.[32] And from 1828–39, he held the Lucasian Professorship of Mathematics at Cambridge, the prestigious chair that Isaac Newton once held. Herschel and Whewell were reportedly the more influential scientists of their generation, but in relation to computing, Babbage's contribution was seminal; he was the first person to design a sequence-controlled computer, his Analytical Engine.

Babbage produced many drawings, but they are all two-dimensional. The most evocative visual representations of his Analytical Engine, and in particular its scale, are those produced by Sydney Padua, for her graphic novel *The thrilling adventures of Lovelace and Babbage*,[33] and also the image created by her for this paper, which is reproduced within Figure 2. His Analytical Engine was designed in four principal parts: the arithmetic unit, storage, input and output, and the sequencing mechanism. The arithmetic unit, which he called the 'mill', was surrounded by the input, the output and the sequencing mechanism. The mill was built like a sponge cake, each layer capable of representing and acting on one digit of a decimal number. Babbage planned to have 50-digit numbers, so there were 50 layers, stacked one above the other. Inside, there was a large circular ring gear that engaged with the little gears around the outside that did the calculation, controlled it, and did input and output. The ring gear was also replicated per digit. Calculations were done in the mill based on addition. The desired sequence of operations and the relevant input of numbers were defined via perforated cards based on those used in Jacquard weaving looms.[34] The storage was also layered, and in columns representing 50-digit numbers. In Babbage's *circa* 1843 plans each

---

[29] Babbage had a number of associations with Ireland, see: *Irish interactions with Charles Babbage regarding his Difference Engines and Analytical Engine.* Available at: https://www.scss.tcd.ie/SCSSTreasuresCatalog/miscellany/TCD-SCSS-X.20121208.001/TCD-SCSS-X.20121208.001.pdf

[30] Charles Babbage, *Table of the logarithms of the natural numbers from 1 to 108000* (London, 1827).

[31] Lovelace, 'Sketch of the analytical engine'.

[32] Charles Babbage, *The Ninth Bridgewater Treatise – a fragment* (London, 1837). Available at: https://victorianweb.org/science/science_texts/bridgewater/intro.htm

[33] Sydney Padua, *The thrilling adventures of Lovelace and Babbage: the (mostly) true story of the first computer* (London, 2016).

[34] James Essinger, *Jacquard's web: how a hand-loom led to the birth of the information age* (Oxford, 2007).

Brian Coghlan et al.

column stored two numbers.[35] Ada Lovelace's 1843 monograph states: 'In the Analytical Engine there would be many more of these columns, probably at least two hundred'.[36] For each number there were 50 little gears, each capable of rotating so as to represent a decimal digit. In the *c.*1843 plans the storage extended away from the mill, and added considerably to the footprint of the engine.

Babbage's planned engine was entirely mechanical, and would have needed precision engineering (but contemporary precision may have been sufficient).[37] It was a very novel and ambitious concept. It is worth quoting his son Henry's words: 'it is to be noted that the engine is designed for analytical purposes, and it would be like using the steam hammer to crush the nut, to use the Analytical Engine to solve common sums in arithmetic'.[38] The vast expense and physical practicalities resulted in only a small portion of the Analytical Engine ever being built. Henry continued: 'I believe that the present state of the design would admit of the engine being executed in metal...I see no hope of any Analytical Engine, however useful it might be, bringing any profit to its constructor...The History of Babbage's Calculating Machines is sufficient to damp the ardour of a dozen enthusiasts'.

In the subsequent Victorian era, others constructed machines that are relevant to the history of computing. From 1837–43 the Swedish father and son Scheutz constructed the first of several difference engines, and a small number of other inventors followed their example.[39] In 1870 at the Royal Society, Jevons exhibited his 'Logic Piano' performing logical inferences.[40] In 1884, Hollerith filed a patent for compiling statistics using punched cards,[41] the basis for the tabulating machines supplied by his company (a predecessor of IBM) for the 1890 census of the USA. But none had the scope or the capability of Babbage's machine, which Lovelace stated: 'holds a position wholly of its own'.[42] It is into this very mechanical and increasingly mathematical scientific era that Percy Ludgate was born.

---

[35] Tim Robinson, Personal communications with Brian Coghlan and Brian Randell (December 2020).

[36] Lovelace, 'Sketch of the analytical engine', 701.

[37] Doron Swade, *The cogwheel brain: Charles Babbage and the quest to build the first computer* (London, 2000). In 1991 the London Science Museum built Babbage's Difference Engine No.2 from its 1846–9 drawings, using only contemporary engineering practices and tolerances.

[38] Major-General H. P. Babbage, 'The Analytical Engine', *Proceedings of the British Association* (paper read at Bath, 12 September 1888). Also included in: H. P. Babbage (ed.), *Babbage's calculating engines: being a collection of papers relating to them, their history, and construction* (London, 1889), 331–8. Available at: https://www.fourmilab.ch/babbage/hpb.html

[39] Michael R. Williams, 'The difference engines', *The Computer Journal* 19:1 (1976), 82–9.

[40] William Stanley Jevons, 'On the mechanical performance of Logical Inference', *Philosophical Transactions of the Royal Society* 160 (1870), 497–518. Available at: https://royalsocietypublishing.org/doi/pdf/10.1098/rstl.1870.0022

[41] Herman Hollerith, 'An electrical tabulating system', *School of Mines Quarterly* (Columbia University) 10:3 (1889), 238–55. Reprinted in Brian Randell (ed.), *The origins of digital computers: selected papers* (Berlin, 1982).

[42] Lovelace, 'Sketch of the analytical engine', 697.

Percy Ludgate (1883–1922), Ireland's first computer designer

Percy Ludgate's family and early life

Percy's father Michael Edward Ludgate (1839–1923),[43] who was born in Kilshannig, joined the North Cork Militia in 1857 with two other members of the Ludgate family.[44] In 1858, the militia was relocated to Kent where the Ludgates transferred to the 21st Regiment of Foot in the regular army. Within two weeks Michael was promoted to Corporal and one year later to Sergeant Musketry Instructor. He married Mary Ann McMahon (1840–1936) in 1863 at Winchester Barracks. Mary was born in Iden, Sussex, but also was from an Anglo-Irish military family, with roots in counties Antrim, Armagh and Suffolk, the Suffolk branch being navy rather than army. Michael served a typical military life at a variety of locations, including India (taking his family with him), Aden, Hythe, Chatham, Gravesend and Winchester. He was ultimately assigned to the 60th Regiment of Foot as a Master Sergeant Instructor, attached to the School of Musketry,[45] an army corps responsible for testing small arms including automatic weapons. In 1876, he retired to Skibbereen, Co. Cork, where Percy's brother Alfred was born in 1881. Michael's service record shows his pension being collected there in 1882, and late in that year he advertised, in the *Skibbereen Eagle*, his services as an after-hours shorthand tutor.[46] It is possible he learned shorthand in the army, since his service record shows his profession as a clerk. He and Mary had at least eight children, only five of whom survived: Thomas (b.1865, Winchester), Augusta (b.1871, Bellary, India), Frederick (b.1875, Gravesend), Alfred (b.1881, Skibbereen), and the eighth and youngest child, Percy, who was born on 2 August 1883 at Townshend Street, Skibbereen (see Pl. II). After that there is a period of seven years during which the family's whereabouts and livelihoods largely elude discovery. Only the whereabouts of Percy's brother Thomas have been established, and only late in this period. In 1888, Thomas married Bedelia Bridget Buckley at St Peter and St Paul's Church in Cork. They settled at Roseville, 80 Sunday's Well, a fine two storey house in Cork city on the east side of the River Lee (opposite the gaol), where they remained until at least 1893.[47]

---

[43] For details of service records, careers, births, marriages and deaths referred to in Section II, see: 'Percy E. Ludgate Prize in Computer Science' (see n. 22), especially sections: *Ancestors of Percy Edwin Ludgate*; *Ancestors of Barbara Hopkins*; *Ancestors of Eileen Mary Ludgate.*

[44] Major J. Douglas Mercer. *Record of the North Cork regiment of militia, with sketches extracted from history of the times in which its services were required, from 1793 to 1880* (Dublin, 1886).

[45] Renamed 'Corps of the Small Arms School' in 1919.

[46] 'M.E. Ludgate', *Skibbereen Eagle*, 16 September 1882, 3; 'M.E. Ludgate', *Skibbereen Eagle*, 18 November 1882, 4.

[47] Bedelia Bridget Ludgate's bridesmaid was her sister Cecilia, who married six months later to John Buckley (same surname) and settled nearby at Vista Villa, 124 Sunday's Well. Cecilia gave birth to two children, Daniel b.1891 and Cecilia b.1893, but when she died at her home after the latter's birth, her death was registered by 'Bridget Ludgate sister present at death Rose Vill.S.Well', so Bedelia still lived at Rose Ville (80 Sunday's Well) in 1893.

Coghlan et al, Proceedings of the Royal Irish Academy: Archaeology, Culture, History, Literature, Volume 121C, 2021. © 2021, RIA, reproduced with permission

Brian Coghlan et al.

PL. II—Percy Ludgate's birthplace, Skibbereen, showing the memorial to the 1798 Rebellion (Image courtesy of The Southern Star).

Michael Ludgate and the remainder of his family are next recorded in *Thom's Directory* at 28 Foster Terrace, Dublin, firstly in 1890 as 'Ludgate, Michael Edward, teacher of shorthand', then in 1891 as 'Ludgate, Michael Edward, teacher of shorthand, Ludgate, Fred, teacher of shorthand, Ludgate, Miss Augusta, teacher of shorthand', and from 1892–8 as 'Ludgate, Michael Edward, sons & daugh, teachers of shorthand', indicating the evolution of a small-scale family business.[48] It had been thought that 28 Foster Terrace no longer existed, but convincing evidence has been found that No. 28 was renumbered in the twentieth century as either No. 46 or No. 47,[49] both of which still exist.

It is known from school records that at age seven to eight years Percy attended St George's Infants School.[50] Randell's 1971 paper states that from eight

---

[48] '28 Foster Terrace, Ballybough, Dublin', in *Thom's Directory* (1890–98).

[49] Gerry Kelly, '46 Foster Terrace' (9 November 2019) and 'Evidence 28 Foster Place was renumbered' (25 November 2020).

[50] NAI, Ireland National School Registers, Roll no. 11624. St George's Parish, St George's Infants School attendance records 1890 and 1891. Percy Ludgate enrolled in Class I on 15 September 1890 (register entry 559) for the year ending 31 March 1891, then is shown as enrolled in Class III for the year ending 31 March 1892, and as leaving the infants school in May 1892.

P∟. III—St George's Church, Temple Street, Dublin, the church that Percy Ludgate attended (Image L_ROY_0072 courtesy of the National Library of Ireland).

to twelve he is said to have attended St George's National School. His family were members of the Church of Ireland,[51] and Randell's 1971 paper also states Percy attended St George's Church, Temple Street, see Plate III.

Percy presumably attended an as yet unidentified secondary school, as attendance was compulsory until age fourteen. In 1898, when he was aged fifteen years, *The London Gazette* published that he was appointed a 'Boy Copyist' in the Irish civil service, then a temporary post for boys aged fifteen to twenty years

---

[51] NAI, 1901 Census returns.

Coghlan et al, Proceedings of the Royal Irish Academy: Archaeology, Culture, History, Literature, Volume 121C, 2021. © 2021, RIA, reproduced with permission

Percy Ludgate (1883-1922), Ireland's First Computer Designer

Brian Coghlan et al.

old.[52] In 1899, his family moved to 30 Dargle Road, Drumcondra (see Pl. IV), in what was a new development on Dublin's outskirts, one mile north-west of Foster Terrace.[53] In that year his father Michael, giving an address in Balbriggan, was imprisoned in Kilmainham Gaol for non-payment of debt (he had to choose between prison or payment of £5.2s, and opted to serve a short sentence as Prisoner 1076 from 8 September to 19 October 1899).[54] The following year his brother Frederick married Alice Walshe.

The 1901 Irish census lists Percy, his mother and a brother, Alfred, at 30 Dargle Road. Percy is listed as a 'Civil Servant (Boy Copyist)' in the National Education Office, while Alfred was a 'commercial clerk (engineering trade)'. Frederick and Alice are listed at 24 Dargle Road (just six doors away), with Frederick as a 'commercial traveller (chemicals)'. His father Michael is listed at 14 Quay Street, Balbriggan, seventeen miles away, perhaps suggesting a divorce, or at least a family break up. Meanwhile, his other brother Thomas and wife Bedelia had moved to 13 Mardyke, on the opposite side of the River Lee in Cork city, with Thomas listed as a land agent's cashier. His sister Augusta is recorded in the 1901 UK census as a 'Deaconess/Missionary' in Liverpool.[55]

In March 1903 Percy was the top Irish candidate in the civil service examinations for assistant clerkship (abstractors).[56] At that time, the civil service arguably preferred classicists.[57] His ability at mathematics may not then have been of great weight, but the fact that he came first in Ireland in the civil service exams was proof he was also strong in classics. He passed the medical, but was not appointed to a position, while at minimum the next six ranked candidates from Dublin have been found to have been appointed to a civil service position.[58] The reason for this rather strange outcome is not yet known.

In August 1903 on his twentieth birthday, Percy's boy copyist post expired, and his short civil service career came to an end.[59] In the same year he seemingly began work on his Analytical Machine, since his 1909 paper, which was submitted in December 1908, states: 'I purpose to give in this paper a short

---

[52] *The London Gazette*, 22 May 1896, 3066–8; 13 July 1897, 3900–01; 14 January 1898; 4 November 1898, 6454–5; 19 September 1902, 6029; 7 November 1902, 7095; 17 March 1903, 1779; 23 August 1904, 5419–20 collectively are relevant to Percy Ludgate's actual and potential career in the civil service.

[53] '30 Dargle Road, Drumcondra, Dublin', in *Thom's Directory* (1899–1936).

[54] 'Percy E. Ludgate Prize in Computer Science' (see n. 22), section *Ancestors of Percy Edwin Ludgate*.

[55] The National Archives (TNA), 1901 UK Census returns.

[56] 'London correspondence', *Weekly Irish Times*, 21 March 1903, 13. Percy Ludgate was ranked the top Irish candidate, and was ranked nineteen in the UK The next six Irish candidates were ranked 22, 25, 29, 33, 35 and 46. All came from Dublin. Only the rankings of the top 50 candidates were published.

[57] Charles Percy Snow, *The Rede lecture (1959): The two cultures* (Cambridge, 1993).

[58] Gerry Kelly, 'Evidence Percy Edwin Ludgate was passed over' (23 November 2020).

[59] Gerry Kelly, 'Percy Edwin Ludgate – Irish civil service' (11 March 2013).

Percy Ludgate (1883–1922), Ireland's first computer designer

PL. IV—30 Dargle Road, Drumcondra, where Percy Ludgate did his work on his Analytical Machine (Image courtesy of Michael Mongan).

account of the result of about six years' work'. It is not known why he became interested in mathematical calculations, let alone in automating them. Randell's 1971 paper stated that: 'It seems almost certain that his work on the analytical machine was a private hobby which, according to his niece, "he used to work at nightly, until the small hours of the morning".'

In October 1904 Percy Ludgate passed the civil service exams for second-division clerkships,[60] but failed the medical, and therefore was not appointed to a position. In February 1905, the 'Case of Mr. Percy Ludgate – Irish Civil Service' was raised in the House of Commons, Westminster, during Questions in the House. The petition was for a new medical examination with a view to his being certified for one or other of the positions for which he had passed exams. He got a negative response from the Financial Secretary of the Treasury, Victor Cavendish MP: 'As nearly a year had elapsed since Mr. Ludgate's medical examination for an assistant clerkship, it was necessary to re-examine him before issuing him a certificate for a second-division clerkship. The result of the medical examination proving unsatisfactory the Civil Service Commissioners were

---

[60] Second-division is the next grade. The examination was announced in *The London Gazette* (23 August 1904), 5419–20. The result is given in n. 61 below.

Brian Coghlan et al.

unable to grant a certificate for either position. The medical requirements are practically the same in both cases'.[61]

The MP for Drumcondra was John Joseph Clancy,[62] a highly educated nationalist politician and King's Counsel (KC), but the MP that took Ludgate's case, Timothy Charles Harrington,[63] was MP for Dublin Harbour. He was Lord Mayor of Dublin in 1901–4, owner of *United Ireland* and *Kerry Sentinel* newspapers, a Corkman and member of the 'Bantry band' of prominent nationalist politicians, and also a highly educated KC.[64] That he took the case of a non-constituent denied a Civil Service post after a failed medical is unusual. It is possible that there was a prior relationship, as Percy's uncle William Ludgate,[65] and William's son Robert,[66] worked for a newspaper in Cork, or that it was simply a reaction to the rejection of Percy's apparent capacity for the role.

Percy Ludgate's year of achievement

In December 1908 Ludgate submitted his paper 'On a proposed Analytical Machine' to the Royal Dublin Society, and in April 1909 it was published in their *Scientific Proceedings*. In July 1909 a detailed review of the paper by Prof. Charles Vernon Boys was published in *Nature*.[67] Percy Ludgate's paper was a significant achievement for a mere clerk to a corn merchant. The Royal Dublin Society's Publications Committee included several fellows of the Royal Society and numerous professors, so Ludgate's proposed machine became known to an elite cohort of scientists.

The minutes of the Publications Committee (a function of the Committee on Science) show the review process his 1909 paper went through.[68] The paper's first reviewer, Prof. A.W. Conway,[69] suggested it was sent to Prof. Boys,[70] presumably for

---

[61] *Hansard* 141 (20 February 1905). Available at: https://api.parliament.uk/historic-hansard/commons/1905/feb/20/irish-civil-service-case-of-mr-percy

[62] 'Obituary of J.J. Clancy KC', *Irish Independent*, 26 November 1928, 7.

[63] 'Death of Mr T.C. Harrington, MP', *Kerry Sentinel*, 16 March 1910, 3.

[64] Patrick Maume, 'Harrington, Timothy Charles', *Dictionary of Irish Biography* (9 vols, Cambridge, 2009). Available at: https://www.dib.ie/biography/harrington-timothy-charles-a3816

[65] William Joseph Ludgate, available at: https://www.scoop-database.com/bio/ludgate_william_joseph_1

[66] Robert Willis Ludgate, available at: https://www.scoop-database.com/bio/ludgate_robert_willis

[67] C.V. Boys, 'A new analytical engine', *Nature* 81 (July 1909), 14–15. Also reproduced in Randell's 1971 paper (see n.8).

[68] *Minutes of the Royal Dublin Society's Publications Committee for 1908–1909* (Dublin, 1908–9).

[69] Conway later became the first Chairman of the Dublin Institute for Advanced Studies, President of the Royal Irish Academy (RIA), and President of University College Dublin (UCD). His descendants have recently been located, but his personal papers have not been found.

[70] Boys (commonly known as C.V. Boys) had already been a Fellow of the Royal Society for twenty years, had recently been President of the Röntgen Society, soon became President of the Physical Society, and later was knighted. His descendants have been located, but his personal papers have not been found.

Percy Ludgate (1883–1922), Ireland's first computer designer

F<small>IG</small>. 2—(left) Charles Babbage imagined in front of his Analytical Engine (Image courtesy of Sydney Padua. © Sydney Padua 2021). (right) Percy Ludgate imagined at his Analytical Machine (Image courtesy of the John Gabriel Byrne Computer Science Collection; portrait of Percy Ludgate, image courtesy of Brian Randell. © Brian Coghlan 2020).

a more expert review (Boys was an acknowledged expert on calculating machines). The minutes say, 'Registrar to communicate Prof Boys remarks to the author and ask him to emphasize the points which he considers important', and later 'Read Mr. Percy Ludgate's letter in reply to the suggestion made to him', presumably by Prof. Boys. The paper says at the start 'Communicated by Prof. A.W. Conway', and this phrase appears in notices of the presentation on 23 February 1909,[71] as well as in the annual report of the Committee on Science,[72] so Conway either communicated the notices or gave the presentation; likely the former as he is not

---

[71] 'Communications', *Freeman's Journal*, 23 February 1909, 6; 'R.D. Society Scientific Meeting', *Irish Independent*, 24 February 1909, 6.

[72] *Proceedings of the Royal Dublin Society*, 96 (11 November 1909 to 10 March 1910). The Report of the Committee on Science and its Industrial Applications says: 'following communications were received…23 Feb. 1909…"On a proposed Analytical Machine," by Percy E. Ludgate. (Communicated by Professor A. W. Conway. M.A.).' Possibly non-member's papers were '(Communicated by)' members, as those so listed are exclusively by non-members.

Brian Coghlan et al.

specifically thanked for any such presentation, whereas the paper says: 'I desire to record my indebtedness to Professor C.V. Boys FRS for the assistance which I owe to his kindness in entering into correspondence with me'.

Percy Ludgate is imagined in Figure 2 seated at a table with his desktop-sized Analytical Machine in front of him, alongside Babbage in front of his large engine.

The huge difference in the Babbage and Ludgate designs adds considerable credibility to Ludgate's statement that he did not know of Babbage's ideas in advance. He included paragraphs on Babbage that surround the body of his paper, together with references within the body, but did not state on what date he was first aware of those ideas. To quote from Ludgate's 1909 paper: 'In order to prevent misconception, I must state that my work was not based on Babbage's results—indeed, until after the completion of the first design of my machine, I had no knowledge of his prior efforts in the same direction. On the other hand, I have since been greatly assisted in the more advanced stages of the problem by, and have received valuable suggestions from, the writings of that accomplished scholar'.

Ludgate's 1909 paper clearly shows his design also had four principal parts: the arithmetic unit, storage, input and output, and a sequencing mechanism. Like Babbage's, Ludgate's arithmetic unit had a mill, which performed addition in the same way. But Ludgate introduced a brand new concept, which he called an 'index', to do multiplication based on what Prof. Boys delightfully termed 'Irish logarithms'.[73] Thus the core of his arithmetic unit did not just do additions; it did multiply-accumulation (MAC), i.e., multiplication followed by addition to any previous result in the mill, the first computer arithmetic unit to do so.[74] Ludgate also proposed mechanisms to do division and logarithms, again introducing a new concept, division by convergent series seeded with an estimate from a mechanical table of initial values.

The arithmetic unit could be stepped through a sequence of operations either manually or under automatic control, the latter using perforated 'formula-paper' to define the desired sequence of operations and perforated 'number-paper' to provide the required set of input numbers, while for output a printer could both print on and perforate paper. Ludgate's scheme to control sequencing was closer to that of modern computers than Babbage's, in that each line of perforations in the formula-paper identified the operation and its operands, whereas Babbage's scheme of separate variable (and combination) cards and operation cards was significantly more complicated. Ludgate's scheme also simplified the provision of the crucial facility of changes to sequencing behaviour if defined conditions, e.g., negative or zero mill contents, occurred.

His storage system was based around two concentric cylinders that held numbers in shuttles. In each shuttle there were 21 rods, one for the sign, and one per digit of a twenty-digit number. The rods protruded from the shuttle

---

[73] Brian Coghlan, 'Percy Ludgate's logarithmic indexes' (see n.14).
[74] MAC is important in signal processing, e.g., in radar and astronomy, and more recently in deep artificial intelligence (deep-AI).

between one and ten units. In order to access a number, the shuttle was rotated to align with the Index. If variable C was in an outer shuttle and variable D was in an inner shuttle, then to perform a calculation $C * D$, each of the storage cylinders was rotated to align those shuttles with the index. Then the shuttles were brought forward along races to engage with the index,[75] and the calculation proceeded with a multiply $C * D$ followed by accumulation of partial products in the mill, following which the shuttles returned to their cylinders. The result in the mill could then be written to a new shuttle and stored somewhere else in one or both of the cylinders. This was a very novel form of storage, and completely new. Ludgate did not discuss the methods of selecting a shuttle, and neither Babbage nor Ludgate discussed the need for, let alone the methods of, selecting a shuttle based on the result of prior calculations.[76]

Only a few machine features are described in Ludgate's 1909 paper, almost everything about its construction is unknown. The 'many drawings of the machine and its parts' that Ludgate states he prepared appear not to have survived (or if they have, they have not been located). But two recently identified contemporaneous articles, one in *English Mechanic and World of Science*,[77] derived from the other in *Engineering*,[78] are of significant assistance in this respect. They contain the first known diagram illustrating the operation of the index. Figure 3 shows an annotated version of this diagram. The text of the articles appeared at first sight to contradict Ludgate's 1909 paper, but subsequent analysis suggested that the text, and the diagram, were probably provided by Ludgate; in fact treating Ludgate's paper and these articles (with their diagram) as equally valid reveal details of the probable working of the index that would have been very hard to arrive at with only Ludgate's 1909 paper.[79]

The index employed logarithmic index 'slides'. The profiles of these slides, portrayed in Figure 4,[80] were such that they obey a logarithmic law that enables them to be used to convert multiplication into addition, i.e., $\log(j*k) = \log(j) + \log(k)$. By means of these slides the decimal digits of numbers were represented in a logarithmic form as lengths, so that multiplication could be

---

[75] A 'race' is a groove or guide along which an object (typically ball or roller bearings, in this case a shuttle) can be moved.

[76] 'Dynamic' selection in modern terminology, considered to be a key requirement for a general-purpose computer.

[77] Engineering (attribution), 'A proposed analytical machine', *English Mechanic and World of Science* 90:2319 (1909/10, 3 September 1909), 111. Discovered by Ralf Buelow, and image provided by Eric Hutton, 19 December 2019. Reproduced in Coghlan, 'Percy Ludgate's analytical machine'.

[78] 'A proposed analytical machine', *Engineering* (20 August 1909), 256–7. Both discovery and provision of the image by Jade Ward, University of Leeds Library, on 14 January 2020. Reproduced in Coghlan, 'Percy Ludgate's analytical machine'.

[79] Coghlan, 'Percy Ludgate's analytical machine'.

[80] Presumed shape based on: David McQuillan, 'The feasibility of Ludgate's analytical machine'. Available at: http://www.fano.co.uk/ludgate/Ludgate.html

Brian Coghlan et al.

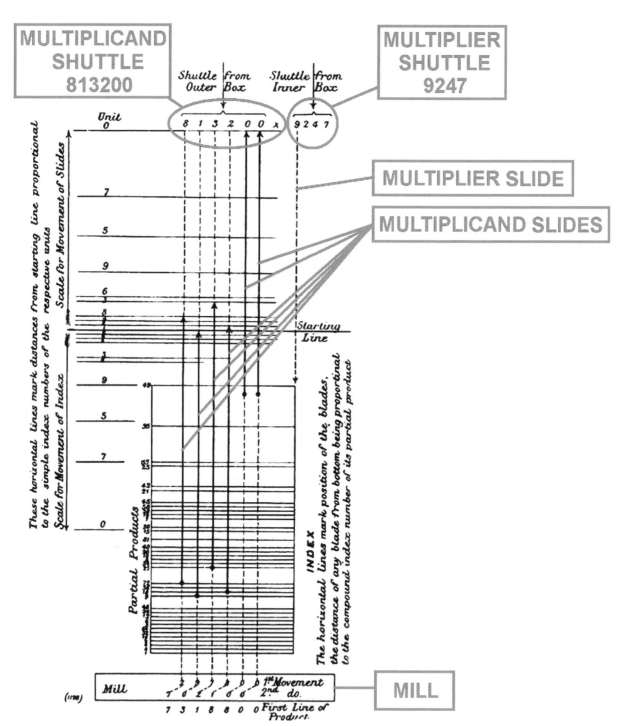

FIG. 3—Annotated version of the diagram from Engineering, 20 August 1909, pp. 256–7 (Image courtesy of the John Gabriel Byrne Computer Science Collection. © Brian Coghlan 2020).

Percy Ludgate (1883–1922), Ireland's first computer designer

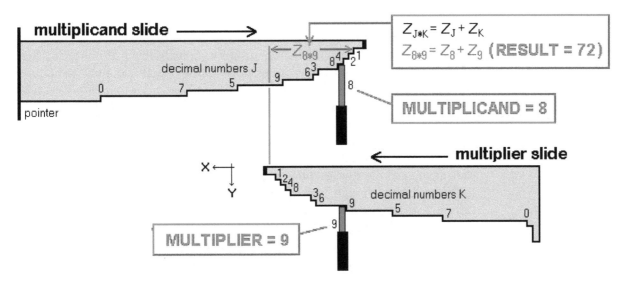

FIG. 4—Multiplication using Ludgate's logarithmic slides for calculation of 8*9 = 72 (Image courtesy of the John Gabriel Byrne Computer Science Collection. © Brian Coghlan 2020).

achieved by addition using simple linear motion. There were twenty identical multiplicand slides, one for each decimal digit. There was just one (opposing) slide for one digit of the multiplier, so the multiplication took place iteratively (one multiplier digit per iteration) with twenty multiplicand digits being simultaneously multiplied by that one multiplier digit.[81] The protruding rods, representing the digits to be multiplied, governed the positions into which the opposing slides could be moved, causing the ends of the slides to increase in overlap from their starting position. The extent of the increase in overlap constitutes the 'logarithm' that represents the resulting two-digit partial product. This digital 'logarithmic' mechanism is therefore very different from a conventional logarithmic slide rule; they do have the addition of logarithmic quantities (lengths) in common, but the logarithms are of a very different form.

Figure 4 shows one of the set of multiplicand slides that has been moved to the right until this particular slide has been stopped by a rod whose vertical extent represents the Irish logarithm of the digit '8'. This is in fact the first digit of the four-digit example multiplicand ('8132') in Ludgate's 1909 paper (the *Engineering* article has a six-digit multiplicand '813200'). Figure 4 also shows that the (single) multiplier slide has been moved to the left until stopped by a rod representing the Irish logarithm of the digit '9' (the first digit of the four-digit example multiplier '9247' of both Ludgate's 1909 paper and the *Engineering* article). However, just how the extent of the increase in overlap is transmitted to the mill is unclear.

---

[81] Coghlan *et al.*, 'Investigating the work and life of Percy Ludgate'.

Brian Coghlan et al.

Ludgate's planned machine was, like Babbage's, to be an entirely mechanical device, but Ludgate's design was evidently very different to Babbage's. It is not known whether the main mechanisms were realisable with the technology of the time. It was small, about the size of a bar fridge. It had a mill to add like Babbage's (but presumably not influenced by him), an index to multiply via Irish logarithms (an entirely new idea), and compact storage via rods and shuttles (another new idea). Sequencing and number input were via a perforated paper sheet or roll in the same vein as Babbage's strung-together punched Jacquard cards, but quite different (and again in all probability not influenced by him). Both the sequencing and the storage differences are at least as significant as the arithmetic differences, even though they are not as great. It is not appropriate here to go into further technical detail, but a much fuller explanation is given in a recent paper,[82] while our most up-to-date understanding is documented in a work-in-progress online paper.[83] There is no evidence that Ludgate ever made any attempt to build this machine. After Ludgate's 1909 paper, and Prof. Boys' 1909 report in *Nature*, there seems to have been very little mention of his work. For example, the 1911 edition of *Encyclopædia Britannica* does not mention his ideas.[84]

No obvious motivations or influences have been discovered, despite scientific, commercial and ancestral searches. In the family domain, for example, there is only a small probability that his brother Frederick could have had much influence on Percy's Analytical Machine design during the seminal period 1903–9, as he lived in Tullamore during 1902–9.[85] Similarly, in the commercial domain, it does not appear that Ludgate's novel multiplier mechanism was influenced by the very different multiplier mechanisms on sale in the late nineteenth century or early twentieth century.[86] And scientifically, Ludgate's multiplier mechanism was totally new, with the only influences likely to be the logarithmic principles developed from the early seventeenth century onwards (again see the work-in-progress online paper).[87]

Percy Ludgate's later life

By 1911, Percy's brother Thomas and wife Bedelia were still in Cork, and his sister Augusta was still in Great Britain.[88] By then also Frederick, Alice and their

---

[82] Coghlan *et al.*, 'Investigating the work and life of Percy Ludgate'.

[83] Coghlan, 'Percy Ludgate's analytical machine'.

[84] Encyclopædia Britannica, *Calculating machines* (1911). Available at: https://en.wikisource.org/wiki/1911_Encyclop%C3%A6dia_Britannica/Calculating_Machines

[85] 'Tullamore Sessions 5 July 1902', *Midland Tribune*, 5 July 1902, 1; 'Tullamore Sessions 1 May 1909', *Leinster Reporter*, 1 May 1909, 3; 'King's County Assizes 2 March 1909', *Midland Tribune*, 6 March 1909, 4.

[86] John Wolff, Personal communication to Brian Coghlan (21 February 2017).

[87] Coghlan, 'Percy Ludgate's analytical machine'.

[88] In the 1911 Irish census Thomas and Bedelia are listed at 173 Gurteenaspig, Bishopstown, Cork. Percy's sister Augusta was recorded in the 1911 UK census as an independent woman boarding in St Pancras, London.

Percy Ludgate (1883–1922), Ireland's first computer designer

daughter Violet had moved to 17 Carlingford Terrace (the street just behind Dargle Road), and Percy, his mother and his brother Alfred were still at 30 Dargle Road.[89] Thomas, Frederick and Alfred were listed in the census respectively as a land agent's cashier, a flour and provision agent, and as a solicitor's clerk. Percy was listed as a commercial clerk to a corn merchant, but it is not known by whom he was employed or when this employment began. It is possible it was a first step towards an accounting career.

In 1914 his second paper, 'Automatic calculating machines', was published in the *Handbook of the Napier Tercentenary Celebration*, which took place in Edinburgh on 24–27 July.[90] The celebration was a major event, which attracted an international audience and had wide coverage in the press. Ludgate's second paper focused on Babbage's Analytical Engine, but briefly mentioned his own 1909 design for his Analytical Machine, and a subsequent and intriguing difference engine design.[91]

Randell's 1971 paper states that 'during the 1914–18 war he [Ludgate] worked for a committee, set up by the War Office, headed by Mr. T. Condren-Flinn, senior partner of Kevans and Son' (accountants, Pl. V),[92] to 'control the production and sale of oats' for the cavalry, which 'involved planning and organisation on a vast scale, and Ludgate was much praised for the major role that he played'.[93] Kevans and Son letterhead of 1917 confirms the senior partners were Edward Kevans and Thomas Condren-Flinn, both chartered accountants.[94] The Ministry of Food established the Food Control Committee for Ireland in September 1917,[95] and it established three oat committees in September 1918,

---

[89] NAI, 1911 Census returns. Note: Carlingford Terrace is now part of Carlingford Road.

[90] Percy E. Ludgate, 'Automatic calculating machines', in, Ellice Martin Horsburgh (ed.), *Napier Tercentenary Celebration: Handbook of the Exhibition* (also published as *Modern instruments and methods of calculation*) (London 1914), 124–7.

[91] Brian Coghlan, 'Speculations on Percy Ludgate's difference engine' (available at n.14).

[92] Kevans and Son occupied Caledonian Chambers, 31 Dame Street, which was designed for the Caledonian Insurance Company in Ruskin-style by the architect James Edward Rogers, see: http://www.patrickcomerford.com/2016/12/a-unique-building-on-dame-street-dublin.html. Kevans and Son was ultimately subsumed into Price Waterhouse Cooper, now known as PwC.

[93] The likely source of this information was 'Mr. E. Dunne…who joined the firm of Kevans & Son in 1921' (from Randell's 1971 paper, see n. 8).

[94] NAI, NAI/PLIC/1 PLIC-1-430, 8 July 1916; PLIC-1-6323, 8 February 1917. Edward Kevans was a founder member, and president (1901–20), of the Irish branch of the Society of Accountants & Auditors (SAA), renamed Society of Incorporated Accountants & Auditors (SIAA) in 1908. Thomas Condren-Flinn was Edward Kevan's clerk, becoming a chartered accountant in 1906, and Kevans' partner and successor.

[95] The Food Control Committee for Ireland (Secretary T.J. Kinnear) was established for 'Defence of the Realm' by the Ministry of Food on 3 September 1917. See: TNA, 'Oats Control Committee', MUN 4/6489, September 1915; TNA, 'War Committee: Oats', CAB-42-14-2, 18 May 1916; TNA, 'War Cabinet: Irish Oats', CAB-24-25-76, 19 August 1917; 'New food regulations: Irish committee appointed', *The Irish Times*, 3 September 1917, 3.

Coghlan et al. Proceedings of the Royal Irish Academy: Archaeology, Culture, History, Literature, Volume 121C, 2021. © 2021, RIA, reproduced with permission

Brian Coghlan et al.

with Kevans and Son coordinating the Midlands Oat Committee.[96] But it has not been possible to locate any actual evidence that the oats were intended for the cavalry, or that Ludgate was praised in relation to this.

Ludgate is often said to have been an accountant, but that was not the case until eight years after his 1909 paper. In 1971, Ludgate's niece stated to Randell that Percy studied accountancy at the Rathmines College of Commerce, where he was awarded a gold medal,[97] and was employed at Kevans and Son. It is not known when the latter began, but certainly in 1917 (aged 34) he was awarded honours in accountancy, [98] after being examined by the Corporation of Accountants.[99] This body held preliminary, intermediate and final examinations in Dublin in June and November, and allowed exemptions from the preliminary examination.[100] Ludgate's name does appear amongst the published 1916 intermediate,[101] and the 1917 final examination results,[102] but preliminary exam results were not published for previous years. Examination by this body provided the least-expensive option for Ludgate to become an accountant in that era,[103] as it did not require putative accountants to be apprenticed,[104] nor to attend university (the subject was not taught in any Dublin university until he was

---

[96] The Food Control Committee for Ireland established Northern, Southern and Midlands Oat Committees in 1918. Kevans and Son coordinated the Midlands Oat Committee. See: 'Notice to oats merchants', *Irish Independent*, 25 September 1918, 4; 'Notice to grain merchants', *Nenagh Guardian*, 26 October 1918, 2.

[97] The college was founded as Rathmines Municipal Technical Institute in 1901, renamed later, and subsumed into Dublin Institute of Technology in 1992 by the *Dublin Institute of Technology Act* (Government of Ireland, 1992). The college records have not been located.

[98] Corporation of Accountants, 'Corporate accountants, results of the June examination', *Freeman's Journal*, 15 September 1917, 2.

[99] The Corporation of Accountants was originally formed in Scotland in 1891, only in 1939 becoming part of the Association of Certified and Corporate Accountants (ACCA), a worldwide body that certifies accountants, then was granted a royal charter in 1974 and changed its name to the Association of Chartered Certified Accountants (again ACCA).

[100] Three examinations had to be passed—preliminary, intermediate and final. Fees for the exams were 10s 6d, £1.1s 0d and £1.11s 6d respectively. The preliminary examination was for the following subjects: Writing from dictation, English grammar and composition, arithmetic, geography, English history, elementary algebra, Euclid book 1, elementary Latin, French or German. Candidates possessing the necessary fundamental knowledge could be exempted from the necessity of passing the preliminary examination.

[101] Corporation of Accountants, 'Corporate Accounts Exams', *Irish Independent*, 2 September 1916, 4; 'Corporate Accountants Examinations', *Freeman's Journal*, 2 September 1916, 3.

[102] Corporation of Accountants, 'Results of the June examination'.

[103] Peter Clarke, 'The historical evolution of accounting practice in Ireland', *Irish Accounting Review* 12 (Special Issue, 2006), 1–23.

[104] Tony Farmar, *The versatile profession: a history of accountancy in Ireland since 1850* (Chartered Accountants Ireland, 2013). Apprenticed accountants are now referred to as 'articled'. The Institute of Chartered Accountants in Ireland (ICAI), now called Chartered Accountants Ireland, operated under a royal charter from 1888, with exacting financial and examination burdens, including the requirement to be apprenticed.

Percy Ludgate (1883–1922), Ireland's first computer designer

Pᴌ. V—31 Dame Street, Dublin (with flag), where Percy Ludgate worked as an accountant (Image courtesy of the John Gabriel Byrne Computer Science Collection).

embarked on his studies).[105] Conceivably these studies and his subsequent oat committee duties and work at Kevans and Son cut short his nocturnal work on his machines.

The first world war ended in November 1918, when Ludgate was 35 years old, overlapped by the Spanish Flu epidemic which lasted from May 1918 to March 1919.[106] The subsequent Irish war of independence (1919–21) led to partition of Ireland in May 1921, and the Anglo-Irish Treaty in December 1921. In that month Ludgate's brother Frederick died of tuberculosis (*phthsis*) at his family's then home at 1 Tolka Villas, Richmond Road, Drumcondra.[107]

Ludgate remained with Kevans and Son as an accountant until his death in October 1922. In Randell's 1971 paper there are two attestations to Ludgate's

---

[105] A Commerce and Accountancy chair was created at University College Galway in 1914, and a Commerce faculty at UCD in 1916.

[106] Ida Milne, *Stacking the coffins, influenza, war and revolution in Ireland 1918–1919* (Manchester, 2020).

[107] Death certificate for Frederick Ludgate (Dublin, 2 December 1921). Available at: http://www.irishgenealogy.ie/

Brian Coghlan et al.

character. Firstly, according to an accountancy colleague he 'possessed characteristics one usually associates with genius…he was so regarded by his colleagues on the staff…humble, courteous, patient and popular'. Secondly, according to his niece Violet, daughter of Frederick and Alice, 'Percy…took long solitary walks'; he was a '…gentle, modest simple man'; she '…never heard him make a condemning remark about anyone', she thought him '…a really good man, highly thought of by anyone who knew him', and he '…always appeared to be thinking deeply'. The only known photograph of him, provided by Violet to Randell, is shown in Pl. VI(a), probably taken in the last five years of his life. He never married.

Randell's 1971 paper stated that in October 1922 Percy developed pneumonia after a holiday in Lucerne and was nursed by Frederick's wife Alice. On 16 October 1922, he died aged 39 at 30 Dargle Road, quickly followed on 22 October 1922 by Alice aged 47 at the Adelaide Hospital.[108] In the UK and Ireland, life expectancy was 57 years in 1922,[109] and his family died at an average age of 58 years, so Percy died younger than expected. That Alice died six days later suggests a highly infectious illness. His death certificate states the cause of death as catarrhal pneumonia,[110] which mostly accompanies diseases like influenza.[111] Both Percy and Alice were buried in the Ludgate family grave in Mount Jerome Cemetery, Harold's Cross, that Percy had purchased for Frederick's burial.[112]

Ludgate's original handwritten will of 26 June 1917 survives.[113] It is the only known example of his handwriting, extracts of which are shown in Plate VI(b). He bequeathed his estate to his mother Mary (who was then aged 82), or, if she predeceased him, to his niece Violet, and made his brother Alfred his executor with £50 in lieu. Alfred administered probate, declaring relatively modest assets that include War Loans and War Savings with interest, bonds in the UK, Post Office and bank accounts plus cash, but just an estimated £10 of personal effects, and no real property, for a total value of £885.7s 4d, including London assets of £192. The debts included medical expenses for a local Drumcondra pharmacy, and a local doctor, Thomas Codd,[114] and to another doctor in Merrion Square, Michael Cox,[115] and funeral expenses plus cemetery fees. Also

---

[108] Death certificates for Percy Edwin Ludgate (Dublin, 16 October 1922); Alice Emily Ludgate (Dublin, 22 October 1922). Available at: http://www.irishgenealogy.ie/

[109] Max Roser, *Life expectancy.* Available at: https://ourworldindata.org/life-expectancy/

[110] Now called 'bronchopneumonia'.

[111] Guthrie McConnell, *A manual of pathology* (Philadelphia, 1915).

[112] Mount Jerome Cemetery, Harold's Cross, Dublin. Available at: http://www.mountjerome.ie/

[113] NAI, NAI/CS/PO/TR Will of Percy Edwin Ludgate, 26 June 1917.

[114] Dr Thomas Paul Codd of 78 Lr. Drumcondra Road, died aged 62 years on 12 Nov 1922 ('self-administered opium poisoning').

[115] Dr Michael Francis Cox (1852–1926), 26 Merrion Square North, was one of Dublin's leading doctors, a physician at St Vincent's Hospital, consulting physician to several Dublin hospitals, including the National Hospital for Consumption, with a large private practice. He was a member of the RIA, elected President of the Royal College of Physicians in 1922, but fell ill in 1923.

Percy Ludgate (1883–1922), Ireland's first computer designer

PL. VI—(a) Percy Ludgate (Image courtesy Brian Randell), (b) Excerpts of Percy Ludgate's handwriting from his will (Images courtesy of National Archives of Ireland).

included is rent of £5.18s 8d to a Miss Lennon, 78 Queen Street.[116] After estate duty including interest was paid, the residue was £812.12s 10d. Probate was proven in Dublin on 23 January 1923 and in London on 12 February 1923.

---

[116] Neither the 1901 or 1911 census lists Lennons at that property (corner of 78 Queen Street and 1 Benburb Street). The 1911 census lists a Lennon family with four daughters aged 12–22 years at 66.3 Queen Street, and another Lennon family with one daughter aged five years at 44.2 Benburb Street, neither close nearby. The 1910 *Thom's Directory* lists a Mary Lennon at 73 Queen Street, but in the 1911 census that building is listed as 'storage'.

Brian Coghlan et al.

Percy Ludgate's final resting place

In 1911 Percy Ludgate's father, Michael Edward Ludgate was living in Omagh, Co. Tyrone, as Edward Ludgate, single and an army pensioner.[117] He applied in 1919 for one of the new state pensions. He died three months after Percy in 1923 in the Union Infirmary, Belfast, and was buried in the Belfast City Cemetery.[118] The burial was arranged and paid for by the Infirmary in the City Cemetery, which suggests he had a pauper's grave.[119]

After Percy's death, it appears that his mother Mary, brother Alfred and niece Violet remained at 30 Dargle Road until 1935. His mother died aged 96 in 1936. When Alfred administered Percy's probate he declared himself an accountant, but no evidence has been found confirming that he was a qualified accountant.[120] After his mother's death Alfred lived in the environs of the North Circular Road. Thomas and Bedelia, whose only daughter died in infancy, moved from Cork to England and settled at Blue Hazel Cottage, Chailey, in the environs of Newhaven and Peacehaven, Sussex.[121] Bedelia died in Newhaven in 1934 and Thomas died in Lewes in 1951. Alfred died in 1953 in the Royal Victoria Eye and Ear Hospital where his niece Violet was employed as an Almoner's clerk, but his death was registered by the hospital, not by Violet. Randell's 1971 paper states that Percy's drawings or manuscript were not found amongst Alfred's effects. Percy's sister Augusta was recorded in the 1939 UK Register as institutionalised in Leavesden Mental Hospital, Watford,[122] where she died in 1954.

Thus by 1954 Percy, his parents and siblings had all died. Those who died in Dublin were buried in the same grave in Mount Jerome Cemetery.[123] His only niece Violet, daughter of Frederick and Alice, lived at 39 Wellington Road, Dublin until 1987, when she died in St Vincent's Hospital in Elm Park. Her probate shows she donated her body to Trinity College Dublin for medical research, finally to be interred at the Cruagh Cemetery, Rathfarnham.[124] Only one person has been found thus far who remembers Violet,[125] and Randell communicated with Violet only by letter. Consequently, until recently it was thought that, since

[117] TNA, 1911 UK Census returns.
[118] For details of births, marriages and deaths referred to in Section V, see: 'Percy E. Ludgate Prize in Computer Science' (see n. 22), especially sections: *Ancestors of Percy Edwin Ludgate; Ancestors of Barbara Hopkins; Ancestors of Eileen Mary Ludgate.*
[119] Belfast City Cemetery, 'Application for internment 58375'.
[120] Alfred Ludgate declared he was an accountant employed at George Drevar Fottrell and Sons (solicitors), 46 Fleet Street, Dublin. There is a letter from his employer attached to probate, but no evidence that he was an accountant. *Thom's Directory* listed accountants, but never listed Alfred Ludgate, whilst continuing to list Percy Ludgate until 1929.
[121] TNA, 1939 UK Register, Thomas Ludgate, Chailey Road, Sussex (1939). They are not in the 1914–15 Worthing Directory, or the 1915, 1922 or 1927 Kelly's Sussex Directory, or the 1934–6 Lewes, Seaford, Newhaven Blue Book.
[122] TNA, 1939 UK Register, Leavesden Mental Hospital, Watford (1939).
[123] Mount Jerome Cemetery (see n. 112).
[124] NAI, 2008/1/6769 Ludgate, Violet: will and associated papers, 1 October 1985.
[125] Brian Coghlan, Personal communications (January 2019).

Percy Ludgate (1883–1922), Ireland's first computer designer

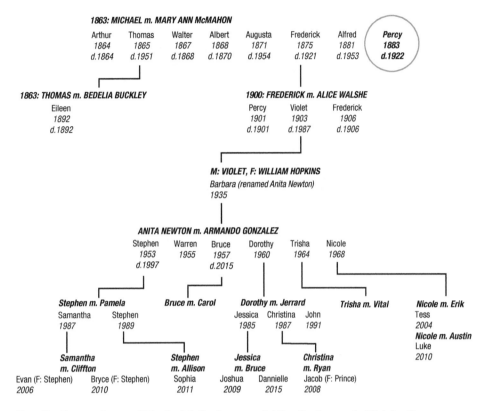

FIG. 5—Descendants of Frederick Ludgate and Alice Ludgate née Walshe (Image courtesy of the John Gabriel Byrne Computer Science Collection and the Gonzalez family. © Brian Coghlan 2020).

Ludgate's parents and siblings had died by 1954, and his niece Violet had died in 1987, that there were no living descendants of his parents. Then in 2018 it was discovered that Violet had given birth to a daughter Barbara in 1935.[126] She was privately adopted and, renamed Anita, brought up in Jamaica. She married and moved to America, where she and her husband raised a family, see Figure 5.

Percy's home at 30 Dargle Road (where he did his work on his Analytical Machine), which was occupied by the Ludgates from 1899–1935, exists in good condition. By contrast the Ludgate grave, occupied by Frederick, Percy, Alice, Mary and Alfred, and owned by Alfred,[127] who made no will, long lay unmarked and in poor condition. Since the grave could only be marked with permission from the owner, or by close descendants, the identification of Anita Newton

---

[126] Brian Coghlan, Personal communications with Trish Gonzalez (7 December 2018), prompted by an old entry (5 August 2013) on a genealogy message board.

[127] National Archives or Ireland, NAI/CS/PO/TR Will of Percy Edwin Ludgate. (see n. 113), and Mount Jerome Cemetery (n. 112).

Brian Coghlan et al.

PL. VII—Ludgate grave, before and after September 2019 (Images courtesy of the John Gabriel Byrne Computer Science Collection).

has allowed this to be addressed, and her children, who are direct descendants of three occupants of the Ludgate grave, were in 2019 allowed to erect a grave marker (Pl. VII).

Concluding summary

Percy Ludgate, who was born, lived, worked and died in Ireland, published his Analytical Machine paper in the *Scientific Proceedings of the Royal Dublin Society* in 1909. Prior to the present paper, he presented two sets of unknowns, one in regard to his life story and the other regarding his machine. Firstly, he died young and single, and by 1987 there were no known descendants of his parents or extended family, so family and personal details were scarce. Secondly, his 1909 paper explained key principles of his machine, but it was not very forthcoming otherwise, as none of the many drawings of the machine and its parts that Ludgate stated in 1909 and 1914 he had prepared had been found. It is possible that the first world war interrupted the dissemination of his work. His ideas seem to have been forgotten by the time of Baxendall's 1926 *Science Museum catalogue of calculating machines and instruments*,[128] in which Ludgate's mention as the designer of a difference engine (rather than of his Analytical Machine, which went unmentioned) stimulated Randell's 1971 investigation.

The new material in the present paper has all been found since 2016, the majority since January 2017. There have been two particularly significant discoveries, firstly at Christmas 2018 of the existence of his niece Violet

---

[128] David Baxandall, *Calculating machines and instruments: catalogue of the collection in the Science Museum* (London, 1926).

Percy Ludgate (1883–1922), Ireland's first computer designer

Ludgate's daughter and descendants, secondly at Christmas 2019 of the article and especially the diagram in an August 1909 issue of the little-known journal *Engineering*.

The first discovery occurred because of the extensive genealogical investigation that has permitted the identification of Percy Ludgate's ancestors back to the 1700s. During these activities it was found that another genealogist was also researching the Ludgate family. This led to the discovery of an unsuspected set of Violet Ludgate's descendants in America (one of whom joined the investigation), who were legally entitled to place a memorial on Percy Ludgate's unmarked grave. His life still presents many mysteries about his education, health, civil service and subsequent career. There remains a hope of finding further descendants of the later generations that were contemporary with Percy Ludgate, and hence the possibility of obtaining through them additional information. But the existing results are a clear testament, firstly to the potential of the Internet and pervasive computing for collaborative historical investigations, secondly, to the power of modern genealogy not only to create extended family trees but also to discover records that unveil the life of a family, and thirdly to the immense value of the continuing digitisation of historical documents and records.

The second discovery was a direct outcome of the investigation's outreach activities, in co-operation with the Ludgate Hub. Prompted by this, Ralf Buelow, of Heinz Nixdorf MuseumForum, found a brief account of the kernel of Ludgate's design, published just a few months after his 1909 paper. Astonishingly this account, like the more comprehensive account in the journal *Engineering* upon which it was based, contained an explanatory diagram that must have been provided by Percy Ludgate himself. This single diagram is of significant value in obtaining a more complete understanding of Ludgate's highly innovative 'Irish logarithms' multiplication mechanism. His machine still presents many mysteries, not least his inspiration for and motivation in pursuing the design of his machine. He designed a potentially disruptive mechanical technology; his storage scheme was extremely compact; his index would have dramatically reduced average multiplication times relative to repeated addition; these and his approach to division were novel. The latter is now common, as is multiply-accumulation (MAC), and grouping of an operation with its operands; these aspects of his design may yet have future application to miniaturised mechanical machines. The new details deduced by this research are a clear testament to the value and ability of even just a single drawing to illuminate hidden facts about a machine, hence the discovery of any further diagrams would be likely to have a significant impact.

The investigation has amassed a very considerable archive,[129] which is, as far as is permitted, available online within the John Gabriel Byrne Computer

---

[129] All the biographical material and genealogical evidence is aggregated in: 'Percy E. Ludgate prize in Computer Science', see note 22. All the technical and other material is available at the Ludgate folder, see note14.

Brian Coghlan et al.

Science Collection. It is hoped that the present paper, and the associated archive, will prompt and encourage further research, especially in Ireland, into all aspects of Percy Ludgate's life and work.

Acknowledgments The authors wish to thank Gerry Kelly (a pseudonymous contributor of detail on Ludgate's previously unknown civil service career and on the history of 28 Foster Terrace); Prof. John Tucker (University College Swansea, UK, for access to D. Riches' report on Ludgate); Canon Eithne Lynch (Mallow Church of Ireland); Dr Susan Hood (Representative Church Body Library, Dublin); Aisling Lockhart (Trinity College Dublin (TCD) Manuscripts and Archives); Dublin City Library and Archive and the Royal Dublin Society Library (for access to records); Prof. Peter Clarke (for detail of the history of accounting); Tim Robinson (for detail of Babbage's storage); Adrienne Harrington (Ludgate Hub); Victoria and Simon Kingston (West Cork History Festival); Lorcan Clancy (for audio and video recordings); the Irish Government (for its generosity in establishing and populating their genealogy website http://www.irishgenealogy. ie/ with Irish civil and church records); the Gonzalez family (for their support); Ralf Buelow of Heinz Nixdorf MuseumsForum and Eric Hutton (for discovery of the articles in *English Mechanic and the World of Science*); Jade Ward of the University of Leeds Library (for discovery of the article in *Engineering*); Sydney Padua (for creating the image within Fig. 2 and permission to publish it); Michael Mongan, *The Southern Star*; the National Library of Ireland; the Director of the National Archives of Ireland (for permission to publish the images in the above Plates and Figures); and to the academic editor and anonymous reviewers of the *Proceedings of the Royal Irish Academy* for their very useful comments. Finally we extend our thanks for the support of the School of Computer Science and Statistics, TCD, for their support for this work and for the John Gabriel Byrne Computer Science Collection.

Extended Family Trees

IF you know descendants who may have photos/documents/etc relating to Percy Ludgate's extended family, PLEASE contact <coghlan@tcd.ie>

—— denotes unproven

© Brian Coghlan 2022

17xx: MATTHEW LUDGATE married MARY Xxxx
Elizabeth 1735 · David 1736 · Jane 1740 · Anne 1743 · Mary 1746 · Mary 1747 d.1748 · Marg. 1749 · John 1752 · Robert 1755

1741: Robert Farmar m. Elisabeth Boulster
George 1745 · Ellen 1747 · Mary 1750 · John 1753 · Elizabeth 1758 · Margaret 1760

1778: JOHN married ELIZABETH FARMAR
Mary 1779 · Katherine 1781 · Matthew 1784 · Michael 1786 · Robert 1788 · Elizabeth 1790
John 1793 · Jane 1795 · David 1797 · John 1799 · Margaret 1801 · Robert 1802 · Thomas 1804
d.1794

17xx: Xxxx Willis m. Xxxx Xxxx
Suzanna 1814

1793: Felix Reed m. Ann Cropley
Mary 1799 · Felix 1801 · Frances 1811

1837: ROBERT married SUZANNA WILLIS
Thomas 1838 · Michael 1839 · Mary 1842 · Elizabeth 1843 · Robert 1845 · Richard 1847
David 1849 · William 1853

1834: Thomas McMahon (1803-1849) m. Frances Reed
Augusta 1834 · Thomas 1836 · Arthur 1838 · Mary 1840 · Robert 1843

1863: MICHAEL married MARY ANN McMAHON
Arthur 1864 d.1864 · Thomas 1865 · Walter 1867 d.1868 · Albert 1868 d.1870 · Augusta 1871 · Frederick 1875 · Alfred 1881 · Percy 1883

1850: Maurice Buckley married Ellen Desmond
Margaret 1851 · Catherine 1853 · Helena 1860 · Maurice 1861 · Michael 1862 · Denis 1864
Mary 1855 · Bedelia 1856 · Cecilia 1858 · Anna 1865 · Elizabeth 1869 · Emily 1871

1873: William Walshe married Mary McFarland
Mary 1874 · Alice 1875 · Eleanor 1877

1888: Thomas Edward Ludgate married Bedelia Buckley
Eileen Mary 1892 d.1892

1900: Frederick Ludgate married Alice Walshe
Percy 1901 d.1901 · Violet 1903 · Frederick 1906 d.1906
d.1987

1877: William Joseph Ludgate married Mary Alice Ferris
William 1878 · Charles 1879 · Isabella 1881 · Robert 1883 · Joseph 1885 · Arthur 1887 · Edward 1889

m.1xxx Matilda Dunbar 1842-1xxx
m.1xxx MaryAnn Sullivan 1844-1xxx
MaryAnn 1866-1xxx · Fanny 1867-1xxx · Margaret 1869-1xxx · Mary 1874-1xxx
Lily 1876-1xxx · Vivian 1879-1xxx · Daisy 1xxx-1xxx
m.1xxx Elizabeth Xxxx 18xx-1xxx · Frances 18xx-1xxx · Maud 18xx-1xxx

Extended Family Trees

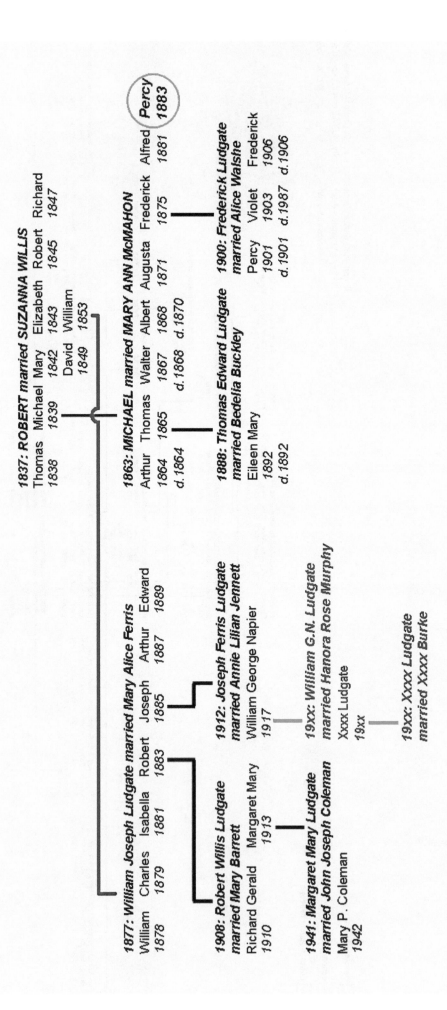

IF you know descendants who may have photos/documents/etc relating to Percy Ludgate's extended family, PLEASE contact <coghlan@tcd.ie>

—— denotes unproven

© Brian Coghlan 2022

1837: ROBERT married SUZANNA WILLIS

Thomas	Michael	Mary	Elizabeth	Robert	Richard
1838	1839	1842	1843	1845	1847

David William
1849 1853

1863: MICHAEL married MARY ANN McMAHON

Arthur	Thomas	Walter	Albert	Augusta	Frederick	Alfred	*Percy*
1864	1865	1867	1868	1871	1875	1881	*1883*
d.1864		d.1868	d.1870				

1888: Thomas Edward Ludgate married Bedelia Buckley

Eileen Mary
1892
d.1892

1900: Frederick Ludgate married Alice Walshe

Percy	Violet	Frederick
1901	1903	1906
d.1901	d.1987	d.1906

1877: William Joseph Ludgate married Mary Alice Ferris

William	Charles	Isabella	Robert	Joseph	Arthur	Edward
1878	1879	1881	1883	1885	1887	1889

1908: Robert Willis Ludgate married Mary Barrett

Richard Gerald   Margaret Mary
1910   1913

1912: Joseph Ferris Ludgate married Annie Lilian Jennett

William George Napier
1917

19xx: William G.N. Ludgate married Hanora Rose Murphy

Xxxx Ludgate
19xx

1941: Margaret Mary Ludgate married John Joseph Coleman

Mary P. Coleman
1942

19xx: Xxxx Ludgate married Xxxx Burke

IF you know descendants who may have photos/documents/etc relating
to Percy Ludgate's extended family, PLEASE contact <coghlan@tcd.ie>

—— denotes unproven

© Brian Coghlan 2022

1xxx: Xxxx McMahon m. Xxxx Yyyy   *1793: Felix Reed [1758-1833] m. Ann Cropley [1771-1xxx]*

Thomas
1803-1849

Mary   Felix   Frances
1799-1xxx 1801-1xxx 1811-1xxx

1834: Thomas McMahon m. Frances Reed

Augusta   Thomas   Arthur   Mary
1834-1890   1836-1xxx   1838-1922   1840-1936

Robert
1843-1xxx

1837: ROBERT married SUZANNA WILLIS

Thomas Michael Mary   Elizabeth Robert Richard David William
1838   1839   1842   1843   1845   1847   1849   1853

m.1xxx   m.1xxx   m.1xxx
Matilda Dunbar   *Mary Ann Sullivan*   *Elizabeth Xxxx*
1842-1xxx   1844-1xxx   18xx-1xxx

MaryAnn   Fanny   Margaret   Mary   Frances   Maud
1866-1xxx   1867-1xxx 1869-1xxx 1874-1xxx 18xx-1xxx 18xx-1xxx

Lily   Vivian   Daisy
1876-1xxx   1879-1xxx 1xxx-1xxx

1863: MICHAEL married MARY ANN McMAHON

Arthur   Thomas Walter Albert Augusta Frederick Alfred (Percy)
1864   1865   1867   1868   1871   1875   1881   1883
d.1864   d.1868 d.1870

1888: Thomas Edward Ludgate   *1900: Frederick Ludgate*
married Bedelia Buckley   *married Alice Walshe*

Eileen Mary   Percy   Violet   Frederick
1892   1901   1903   1906
d.1892   d.1901   d.1987   d.1906

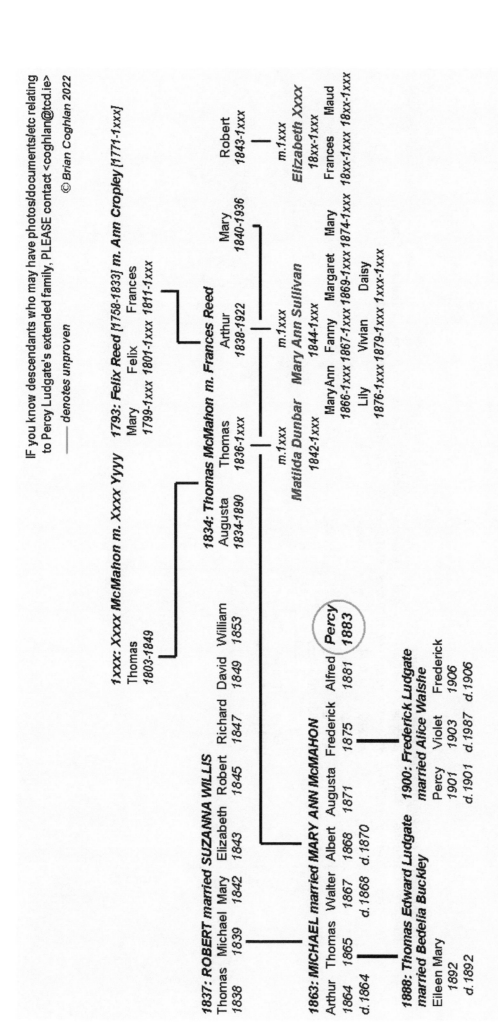

Extended Family Trees

Extended Family Trees

IF you know descendants who may have photos/documents/etc relating to Percy Ludgate's extended family, PLEASE contact <coghlan@tcd.ie>

© Brian Coghlan 2022

—— denotes unproven

1850: Maurice Buckley married Ellen Desmond

Margaret	Catherine	Mary	Bedelia	Cecilia	Helena	Maurice	Michael	Denis	Anna	Elizabeth	Emily
1851-1xxx	1853-1xxx	1855-1xxx	1856-1934	1858-1893	1860-1xxx	1861-1xxx	1862-1xxx	1864-1xxx	1865-1xxx	1869-1xxx	1871-1xxx
m.1898	m.1875		m.1888	m.1888		m.1888					
John Cosgrove ?	John Keyes		Thomas Ludgate 1865-1951	John Buckley 18xx-1xxx		Anne Healy?					

Children of Bedelia & Thomas Ludgate:

Eileen 1892-1892

Daniel 1891-19xx   Cecilia 1893-19xx

IF you know descendants who may have photos/documents/etc relating to Percy Ludgate's extended family, PLEASE contact <coghlan@tcd.ie>

© Brian Coghlan 2022

—— denotes unproven

18xx: William Walshe [18xx-1xxx] married Mary Xxxx [18xx-1xxx]

William
1843-1901

18xx: John McFarland [18xx-1xxx] married Alice Murphy [18xx-1xxx]

Alice	Mary	Margaret	John
1845-1xxx	1851-1xxx	1856-1xxx	1858-1xxx

1873: William Walshe married Mary McFarland

Mary	Alice	Eleanor
1874-1xxx	1875-1922	1877-1xxx

m.1900
Frederick
Ludgate
1875-1921

Percy	Violet	Frederick
1901-1901	1903-1987	1906-1906

with
William
Thaddeus
Hopkins
1897-19xx

Barbara
1935-2xxx

[see RIA paper]

Extended Family Trees

Extended Family Trees

IF you know descendants who may have photos/documents/etc relating to Percy Ludgate's extended family, PLEASE contact <coghlan@tcd.ie>

© Brian Coghlan 2022

—— denotes unproven

1829: Francis Hopkins *[abt.1799-1868] married Bridget Shanly [1806-1836]*

Michael	William	Francis	Catherine
1831-1xxx	1835-1908	1837-1xxx	1840-1xxx

18xx: William Hopkins married Eliza Coleman *[abt.1840-1880]*

Francis	Bridget	Ellen	William	Edward	Elizabeth	Mary Ann	Eliza
1859-1869	1862-1xxx	1864-1xxx	1866-1909	1868-1xxx	1873-1xxx	1876-1xxx	1878-1909

m.1xxx
Thomas Cullen
[1879-1940]

1896: William Hopkins married Catherine Frances Coffey *[1874-1906]* [see Coffey tree]

William Thaddeus	Kathleen Margaret	Gertrude M.	Helen	Ethel
1897-19xx	1898-1xxx	1900-1945	1902-1992	

with
Violet Ludgate
1903-1987

m.19xx
John Patrick Molony
1xxx-1xxx

Stokes
1904-1995

Barbara	Daniel	Catherine
1935-2xxx	1945-2xxx	1949-1xxx
[see RIA paper]		

with
Sylvia Devitt
1944-2010

Nicolette Connolly
1964-2xxx

with
Xxxx Yyyy
1xxx-1xxx

m.19xx
Brendan McPeake
1948-2020

Shane	Jamie	Niamh
1984-2xxx	1988-2xxx	1989-2xxx

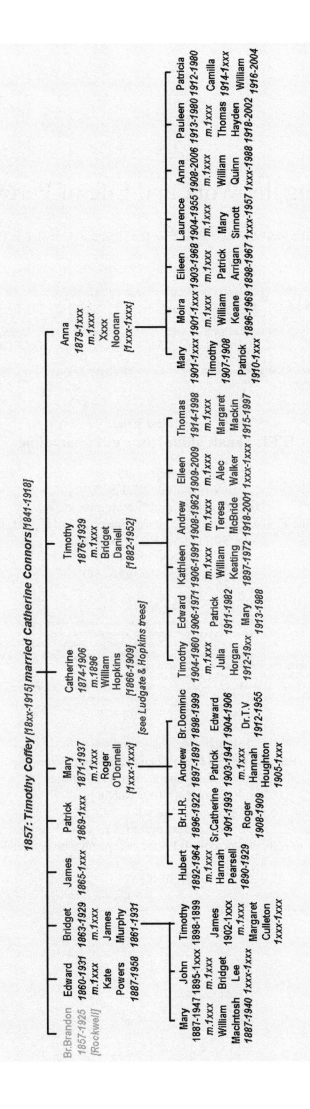

IF you know descendants who may have photos/documents/etc relating to Percy Ludgate's extended family, PLEASE contact <coghlan@tcd.ie>

—— *denotes unproven*

© Brian Coghlan 2022

1857: Timothy Coffey [18xx-1915] married Catherine Connors [1841-1918]

Br.Brandon 1857-1925 [Rockwell]

Edward 1860-1931
m.1xxx
Kate Powers 1887-1958

Bridget 1863-1929
m.1xxx
James Murphy 1861-1931

James 1865-1xxx

Patrick 1869-1xxx

Mary 1871-1937
m.1xxx
Roger O'Donnell [1xxx-1xxx]

Catherine 1874-1906
m.1896
William Hopkins [1866-1909]
[see Ludgate & Hopkins trees]

Timothy 1876-1939
m.1xxx
Bridget Daniell [1882-1952]

Anna 1879-1xxx
m.1xxx
Xxxx Noonan [1xxx-1xxx]

Mary 1887-1947
m.1xxx
William Macintosh 1887-1940

John 1895-1xxx

Timothy 1898-1899

James 1902-1xxx
m.1xxx
Bridget Lee 1xxx-1xxx

Margaret Culleton 1xxx-1xxx

Hubert 1892-1964
m.1xxx
Hannah Pearsell 1890-1929

Br.H.R. 1896-1922

Andrew 1897-1897

Br.Dominic 1898-1999

Sr.Catherine 1901-1993

Patrick 1903-1947
m.1xxx
Hannah Houghton 1905-1xxx

Edward 1904-1906

Roger 1908-1909

Dr.T.V 1912-1955

Timothy 1904-1960

Edward 1906-1971

Kathleen 1906-1991
m.1xxx
Patrick William Keating 1897-1972

Andrew 1908-1962
m.1xxx
Teresa McBride 1918-2001

Eileen 1909-2009
m.1xxx
Alec Walker 1xxx-1xxx

Thomas 1914-1998
m.1xxx
Margaret Mackin 1915-1997

Patrick 1911-1982
m.1xxx
Julia Horgan 1912-19xx

Mary 1913-1888

Mary 1901-1xxx

Moira 1901-1xxx
m.1xxx
William Keane 1896-1969

Eileen 1903-1968
m.1xxx
Patrick Arrigan 1898-1967

Laurence 1904-1955
m.1xxx
Mary Sinnott 1xxx-1957

Anna 1908-2006
m.1xxx
William Quinn 1xxx-1988

Pauleen 1913-1980

Patricia 1912-1980
m.1xxx
Thomas Hayden 1918-2002

Camilla 1914-1xxx

William 1916-2004

Timothy 1907-1908

Patrick 1910-1xxx

Extended Family Trees

Investigating the Work and Life of Percy Ludgate

Brian Coghlan, Brian Randell, Paul Hockie, Trish Gonzalez, David McQuillan, Reddy O'Regan

Abstract: Percy Edwin Ludgate (1883–1922) is notable as the second person to publish a design for an Analytical Machine, the first after Babbage's "Analytical Engine." We outline the initial results of the first new investigation into the work and life of Percy Ludgate since Randell's papers of nearly 50 years ago and nearly 100 years after Ludgate's death. First, we examine the principles of his machine and how it was constructed and worked. Second, we outline his life. We present a range of new material, including two significant discoveries, one concerning Ludgate's machine and the other his family.

Index Terms—**Percy Ludgate, Analytical Machine, Irish Logarithms, Multiply-Accumulate (MAC), Division by Convergent Series, Charles Babbage, Analytical Engine**

Published in the
IEEE Annals of the History of Computing

Page(s): 1 – 19
Date submitted: 27th July 2020
Date of Publication: 17th November 2020
Volume 43, Issue 1, pp. 19-37, January-March 2021, © 2021, IEEE
doi: 10.1109/MAHC.2020.3038431
Print ISSN: 1058-6180
Electronic ISSN: 1934-1547

Authors
Brian Coghlan is a retired Senior Lecturer with the School of Computer Science and Statistics, Trinity College Dublin, The University of Dublin, Ireland. He is curator of The John Gabriel Byrne Computer Science Collection. Contact him at <coghlan@cs.tcd.ie>.
Brian Randell is an Emeritus Professor of Computing Science, and a Senior Research Investigator with the School of Computing, Newcastle University, U.K. His book *The Origins of Digital Computers: Selected Papers* was first published in 1973, by Springer Verlag. Contact him at <Brian.Randell@ncl.ac.uk>.
Paul Hockie is a genealogist from London, U.K. Contact him at <paul@hockie.co.uk>.
Trish Gonzalez is a genealogist from Florida, U.S.A. Contact her at <trishalg03@aol.com>.
David McQuillan is a retired systems designer from the U.K. who gives occasional talks on mathematics. Contact him at <dmcq@fano.co.uk>.
Reddy O'Regan is a retired solicitor from Skibbereen, Co.Cork, Ireland. Contact him at <oreganic@live.ie>.

ARTICLE

Investigating the Work and Life of Percy Ludgate

Brian Coghlan, *Trinity College Dublin, The University of Dublin, Ireland*

Brian Randell, *Newcastle University, U.K.*

Paul Hockie, *Genealogist, U.K.*

Trish Gonzalez, *Genealogist, USA*

David McQuillan, *Retired Systems Designer, U.K.*

Reddy O'Regan, *Retired Solicitor, Ireland*

Percy Edwin Ludgate (1883–1922) is notable as the second person to publish a design for an Analytical Machine, the first after Babbage's "Analytical Engine." We outline the initial results of the first new investigation into the work and life of Percy Ludgate since Randell's papers of nearly 50 years ago and nearly 100 years after Ludgate's death. First, we examine the principles of his machine and how it was constructed and worked. Second, we outline his life. We present a range of new material, including two significant discoveries, one concerning Ludgate's machine and the other his family.

Percy Edwin Ludgate (1883–1922) is notable as the second person to publish a design for an Analytical Machine,[1,2] the first after Babbage's "Analytical Engine."[3] Strangely enough, he was not a scientist, but a clerk to a corn merchant (and subsequently an accountant), born in Skibbereen and employed in Dublin, Ireland. He was working in his spare time, presumably on his own, from 1903 to 1909, and was not aware of Babbage's work until later. Indeed his engine differed greatly from Babbage's in that it was largely based on multiplication using rods in shuttles plus logarithmic "slides" like a digital evocation of slide rules, whereas Babbage's was based on addition using cogs and wheels (interestingly, Babbage's initial difference engine efforts were based on sliding rods, although they were not logarithmic[4]). Both Ludgate and Babbage based input on perforated paper (separately for instructions and data), but Ludgate notably unified instruction and operand sequencing, whereas Babbage separated them. Either machine would be capable of doing everything a modern computer of their scale could do, although very slowly.

Randell's 1971 and 1982 papers[5,6] revealed Ludgate and resurrected him from obscurity, but nothing new has been published on him since then. Prof. J. G. Byrne collected an original offprint of Ludgate's 1909 paper for what became The John Gabriel Byrne Computer Science Collection[7] in the Department of Computer Science, Trinity College Dublin, and then in 1991 instigated a prize in memory of Ludgate.[8] The Collection now holds copies of all the known literature and records relating to Ludgate and, spurred by the cataloging of this collection, since 2016 the collective authors have undertaken a detailed investigation of his work and life. This and Coghlan's work[9] outline the initial results of this first new investigation into the work and life of Ludgate since Randell's papers of nearly 50 years ago and nearly 100 years after Ludgate's death, and provide a range of new material.

Prior to the present investigation, there were two central mysteries about Percy Ludgate. One has to do with his machine and the other to do with his life.

First, his 1909 paper explains key principles of his machine, but said it was: "not possible in a short paper...to go into any detail as to the mechanism...I must therefore, confine myself to a superficial description, touching only points of particular interest or importance." So machine details are scant. However in 1909, he said he had made: "many drawings of the machine

1058-6180 © 2020 IEEE
Digital Object Identifier 10.1109/MAHC.2020.3038431
Date of publication 17 November 2020; date of current version 5 March 2021.

Coghlan et al, IEEE Annals of the History of Computing, Volume 43, Issue 1, pp. 19-37, 2021. © 2021, IEEE, reproduced with permission

Investigating the Work and Life of Percy Ludgate

Investigating the Work and Life of Percy Ludgate

Coghlan et al, IEEE Annals of the History of Computing, Volume 43, Issue 1, pp. 19-37, 2021. © 2021, IEEE, reproduced with permission

ARTICLE

TABLE 1. Publication process of Percy Ludgate's 1909 paper.

Date	Who	Action
Prior to 8-Dec-1908	Percy Ludgate	Submitted his paper.
Tuesday 8-Dec-1908	RDS Science Committee	"Mr. Percy Ludgate's paper on an Analytical Machine was referred to Prof Conway for a report."
Tuesday 12-Jan-1909	RDS Science Committee	"Read Prof Conway's report on Mr Ludgate's paper, it was decided to adopt his suggestion & send it to Prof Boys for his report".
Tuesday 9-Feb-1909	RDS Science Committee	"Mr. C.V. Boys letter on Mr. Ludgate's paper was read & the paper was accepted for the next meeting. The Registrar to communicate Prof Boys remarks to the author & ask him to emphasize the points which he considers important"
Tuesday 23-Feb-1909	Percy Ludgate or Prof.Conway	Presented his paper to the next Scientific Meeting (from the published paper, which says: "Read February 23").
Tuesday 8-Mar-1909	RDS Science Committee	"Read Mr. Percy Ludgate's letter in reply to the suggestion made to him in reference to his paper on a proposed analytical machine." And: "The three papers read at the last meeting by Mr. Brown, Mr. Moss, & Mr. Ludgate were ordered to be printed."

and its parts," and five years later in 1914, he said: "Complete descriptive drawings of the machine exist, as well as a description in manuscript." Clearly, he put a lot of work into this. A machine of the complexity that he designed would take many diagrams. Although with a very small probability, somewhere these diagrams might exist. The ultimate objective of the present investigation is to find those drawings and the manuscript, should they still in fact exist.

Second, Percy Ludgate died young and single in 1922 aged 39. By 1987, there were no other known descendants of his parents. So records are scant too. Randell 1971[5] said: "an archivist stated that by all normal criteria, it was clear that he had never existed." Yet he did exist, with, it turns out, a sizeable number of relatives and acquaintances, potentially with extant descendants. The second objective of the present investigation was and remains to find any such descendants, and any related documents, photographs, and memories that they may have.

PERCY LUDGATE'S ANALYTICAL MACHINE

Percy Ludgate's 1909 Paper

Percy Ludgate was the second person in history to publish a design for an Analytical Machine. His 1909 paper was published in the Scientific Proceedings of the Royal Dublin Society. We have found details in the minutes of the RDS Science Committee[10] that establish the process the paper went through (see Table 1).

Thus, the first reviewer (Prof. Conway) suggested it be sent to Prof. C. V. Boys, presumably for a more expert review. The minutes say "Registrar to communicate Prof. Boys remarks to the author & ask him to emphasize the points which he considers important," and later "Read Mr. Percy Ludgate's letter in reply to the suggestion made to him" (presumably by C. V. Boys).

Then Ludgate says at the start of his paper "I desire to record my indebtedness to Professor C. V. Boys FRS for the assistance which I owe to his kindness in entering into correspondence with me."

Ludgate's paper then includes four paragraphs at the start and two paragraphs at the end on Babbage, plus references to Babbage in the middle of the text. The huge difference in how Babbage and Ludgate envisaged their machines suggests he did not know of Babbage's ideas in advance, and that his paper's reviewer Prof. C. V. Boys' suggestions led to him adding the paragraphs on Babbage surrounding the body of his paper, plus the references to Babbage within the body.

No related material has been found among Prof. C. V. Boys' papers, neither in the London Science Museum nor with his descendants.[11] The paper says at the start "Communicated by Prof. A. W. Conway," and this phrase appears in notices of the presentation on February 23, 1909,[12] so Conway either communicated the notices or gave the presentation; likely the former as he is not specifically thanked. Despite extensive searches, Prof. Conway's papers and descendants have yet to be found.

Ludgate's Analytical Machine

Ludgate's design had four principal parts: the arithmetic unit, storage, input/output, and a sequencer for program control, as illustrated in the conjectural sketch in Figure 1. It can be seen that while Babbage used a cylindrical envelope for calculation and a rectangular envelope for storage, Ludgate did somewhat the reverse.

Like Babbage's, Ludgate's machine also had a Mill (a term he used in his 1909 paper), which did addition in the same way. But notably, Ludgate introduced a brand new concept, which he called an "Index," to do

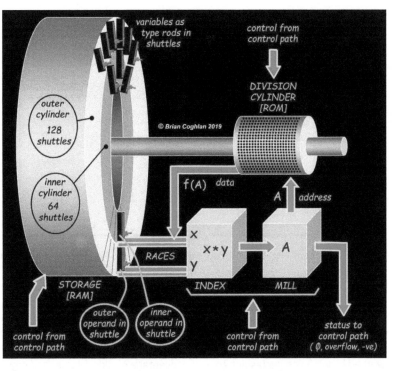

FIGURE 1. Conjectural sketch of Percy Ludgate's Analytical Machine, where "control path," "data path," "status," "RAM," "ROM," "address," "data," are modern terms, all other terms are from Ludgate's 1909 paper. Image reproduced courtesy The John Gabriel Byrne Computer Science Collection.

multiplication based on what are now called Irish Logarithms. The core of his machine did not just do additions; it did multiply-accumulation (MAC), i.e., multiply followed by add to any previous result in the Mill. MAC is important in signal processing, e.g., in radar and astronomy, and more recently in deep AI.[13]

The machine could be operated either manually or under program control, the latter using perforated "formula-paper" for code and "number-paper" for data. He also added some cylindrical mechanisms to do division and logarithms (these cylinders were somewhat akin to the pegged cylinders used in musical boxes and automata, but with rows of holes of differing depths, rather than protruding pegs). Again this introduced a new concept, division by convergent series seeded with an estimate from a mechanical equivalent to read-only memory, and executed under the control of a mechanical equivalent of a built-in subroutine.

A further pièce de résistance was his Storage system, which was based around two concentric cylinders that held numbers in shuttles. Rods protruded from the shuttles. In each shuttle, there were 21 rods, one for the sign, and one per digit of a 20-digit number. The rods protruded from the shuttle between one and ten units, so in principle to represent the value 7 the rod protruded

seven units (but see further below). In order to access a number, the shuttle was rotated to align with the Index. If variable C was in an outer shuttle and variable D was in an inner shuttle, then to perform a calculation $C*D$, each of the storage cylinders was rotated to align those shuttles with the Index. Then, the shuttles were brought forward along the "Races" to engage with the Index, and the calculation proceeded with a multiply $C*D$ followed by accumulation of partial product units and tens in the Mill, then the shuttles returned to their cylinders. The result (in the Mill) could then be written to a new shuttle and stored somewhere else in one or both of the two cylinders. This was a very novel form of random access memory, and completely new.

As stated above, the machine's core operation was multiply-accumulate, processed jointly in his Index and Mill. This employed a novel variant of long multiplication that iterated per digit from the most significant digit of the multiplier toward the least significant digit, i.e., in reverse to the traditional order. For each digit of the multiplier, the units of the partial product were generated by the Index and added to the accumulator by the Mill, then the tens of the partial product were generated and added to the accumulator. Figure 2 shows an example illustrating both the novel iteration order and how the

Coghlan et al, IEEE Annals of the History of Computing, Volume 43, Issue 1, pp. 19-37, 2021. © 2021, IEEE, reproduced with permission

Investigating the Work and Life of Percy Ludgate

ARTICLE

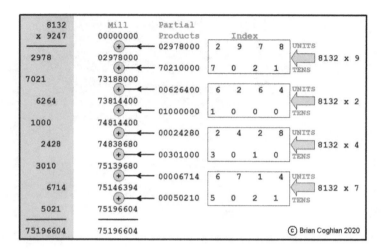

FIGURE 2. Example of Ludgate's variant of long multiplication, calculating 8132 × 9247 (excludes his treatment of carries). Image reproduced courtesy The John Gabriel Byrne Computer Science Collection.

partial products are split into units and tens (but excluding any treatment of Ludgate's carry propagation, which is explored in Coghlan's work[9]).

Ludgate's Index worked in an analogous fashion to a slide rule, which has scales that are more fine as the numbers increase, such that they obey the logarithmic law that they convert multiplication into additions, i.e., $\log(j*k) = \log(j) + \log(k)$. The Index embodied a full single-digit multiplication table (for numbers 1–9), as did a few other calculating machines at this time, such as the Millionaire.[14] However, the numbers were represented in a logarithmic form, and as lengths, rather than angular positions (of a cogwheel), so that multiplication could be achieved by addition, using simple linear motion. These were not ordinary logarithms, where when real numbers increase the difference between their logarithms continuously reduces. Rather, they were integer values that have been referred to as "Irish Logarithms." For two operands Z_J and Z_K, Ludgate's index numbers ensured $Z_Y = Z_{J*K} = Z_J + Z_K$. For example, indexes $Z_3 = 7$ and $Z_5 = 23$, therefore $Z_{15} = Z_{3*5} = Z_3 + Z_5 = 30$. It is not known how Ludgate arrived at his set of logarithms. This matter is discussed in Coghlan's work,[9] showing a method that we can speculate he might have devised and used. As can be seen in Table 2, simple indexes form a nonmonotonic function of the decimal operands, but a monotonic function of the ordinals (because ordinals represent the rank order of simple indexes). Therefore, ordinals may be used as a proxy for the decimal operands in circumstances where monotonicity is desirable, and Ludgate employed this. This is explained further below, and outlined in more detail in Coghlan's work.[9]

Logarithms and slide rules may seem ancient technology, and indeed go back to Napier c.1614[15] and Oughtred c.1622.[16] Slide rules are not commonly used nowadays, but they once were employed widely, especially for scientific and engineering calculations. The advantage of slide rules is their physical simplicity in providing their functionality, compared to other alternatives, even though to many they may now appear a curious oddity. In a similar vein, Ludgate's accumulation worked in an analogous fashion to Napier's "bones," otherwise called "rods," a precursor to slide rules and once widely used across Europe (as an aid to manual calculation) to accumulate partial product units and tens digits that were inscribed on their upper surface.

In summary, Ludgate's machine was entirely mechanical, the main mechanisms seem realizable with the technology of the time, and it was *small* (from dimensions in Ludgate's paper, about the size of a bar fridge). Ludgate's design was *very* different to Babbage's. It had a Mill to add like Babbage's (but presumably not influenced by him), an Index to multiply via Irish Logarithms (a new idea), and compact Storage via rods and shuttles (another new idea). Programming and input/output were via punched paper in the same vein as Babbage's but quite different (and again in all probability not influenced by him). There is no evidence that Ludgate ever made any attempt to build his machine, and Ludgate's drawings had never been found. We are trying to "reimagine" his design, but it is difficult to say if anything beyond the Storage and Index can ever be built.

ASPECTS OF LUDGATE'S MACHINE

Features

Only a few features are described in Ludgate's 1909 paper. The rest must be deduced by contextual

TABLE 2. Ludgate's simple and compound logarithmic indexes (reproduced from Ludgate's 1909 paper, Tables 1 and 2).

Decimal operand	Simple index	Ordinal number
0	50	9
1	0	0
2	1	1
3	7	4
4	2	2
5	23	7
6	8	5
7	33	8
8	3	3
9	14	6

Partial product	Compound index	Partial product	Compound index	Partial product	Compound index
1	0	15	30	36	16
2	1	16	4	40	26
3	7	18	15	42	41
4	2	20	25	45	37
5	23	21	40	48	11
6	8	24	10	49	66
7	33	25	46	54	22
8	3	27	21	56	36
9	14	28	35	63	47
10	24	30	31	64	6
12	9	32	5	72	17
14	34	35	56	81	28

analysis of the paper, largely a process of logical inference or elimination of false propositions, and mechanical or mathematical speculations. They are summarized in Table 3, but explored in Coghlan's work.[9]

Unknowns

Almost everything about its construction is unknown. For a small sample of these unknowns, see Table 4. Many such issues are explored in considerable detail in Coghlan's work.[9]

Reducing the Unknowns: New Discoveries

Clearly, the discovery of drawings by Ludgate would reduce the unknowns, so this has been an ultimate target of the research, but any discovery of new information about Ludgate's design would help. And in fact, prompted by our open call for information about Ludgate, just before Christmas 2019, Ralf Buelow of Heinz Nixdorf MuseumsForum with Eric Hutton[17] discovered a pair of articles about Ludgate's machine in the 1909 issues of the popular science magazine *English Mechanic and World of Science*.[18] One was a

TABLE 3. Features of Percy Ludgate's Analytical Engine (¥ denotes inferences that are detailed in Coghlan's work[9]).

	Base operation is multiply-accumulate (MAC) not addition
	Multiply is done with Irish Logarithms by an INDEX
	Long multiply starts at left digit of multiplier
¥	Numbers must be fixed-point
	Multiply-accumulate partial-products are added units first, then tens, by a MILL
¥	Timing implies pipelining tens carry adds
¥	Instruction set: ADD, SUBTRACT, MULTIPLY, DIVIDE, STORE, CONDITIONAL BRANCH
	Two-operand addressing for load
¥	Two-operand addressing for store
	Fast for 1909: ADD/SUB 3 sec, MUL 10 sec, DIV 90 sec, LOG 120 sec
¥	Storage of 192 variables implies (64 inner + 128 outer) shuttles, equispaced
¥	Hence storage size implies binary storage addressing
	Numbers stored via rod for sign & every digit protruding 1–10 units
	Data input/output via punched number-paper (or upper keyboard)
	Program input/output via punched formula-paper (or lower keyboard), one instruction per row
¥	Manual preemption
¥	Measurements in units of an eighth of an inch
¥	Main shaft gives 'cycle time' of a third of a second
	Small size: estimated by Ludgate as 0.5m H x 0.7m L x 0.6m W

Coghlan et al, IEEE Annals of the History of Computing, Volume 43, Issue 1, pp. 19-37, 2021. © 2021, IEEE, reproduced with permission

ARTICLE

TABLE 4. Sample of known unknowns of Percy Ludgate's Analytical Engine.

How shuttles were selected in storage cylinders	Any internal dimensions
How a shuttle was moved	Any internal timing
How the INDEX mechanism worked	Almost everything about program control
How the MILL mechanism worked	Almost everything about input & output

very brief summary[19] of Ludgate's 1909 paper, with no new information.

The other article[20] did indeed promise new information, if validated, and included the first known diagram of Ludgate's machine! This article was attributed to *Engineering*. A search by Jade Ward of University of Leeds Library discovered the article was in fact an abbreviated extract of an article[21] published a month earlier, i.e., just over three months after Ludgate's original Royal Dublin Society paper, in *Engineering* (a London-based monthly magazine founded in 1865). This article, which from here onwards will be referred to as *Engineering*,[21] is reproduced in full as the Appendix.

Figure 3 shows an annotated version of the diagram from *Engineering*[21] (an identical copy of which appeared in *English Mechanic and World of Science*[20]), which presumably was provided, and perhaps actually drawn, by Ludgate.

The text of the article of *Engineering*[21] at first appeared to us to contradict Ludgate's 1909 paper, but as analysis has progressed it has begun to seem more likely that the text was provided by Ludgate too (the article includes a detailed explanation of the diagram). Ludgate and the writer of the article are quite precise in what they say. In fact, treating Ludgate's paper and the article as equally valid has uncovered ways the design may have been that would have been very hard to arrive at with only Ludgate's 1909 paper.

For example, we have deduced the following essential principles of operation. Ludgate employed

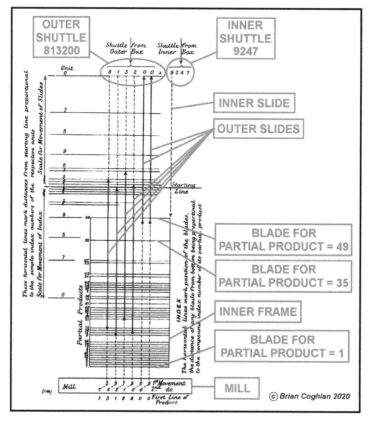

FIGURE 3. Annotated version of the diagram from *Engineering*, August 20, 1909, pp. 256–257.[21] See the Appendix for the original full-size image. Image reproduced courtesy The John Gabriel Byrne Computer Science Collection.

ARTICLE

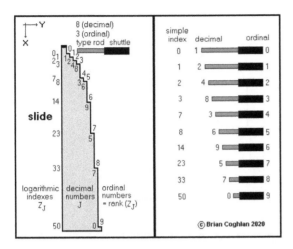

FIGURE 4. (Left) Ludgate's logarithmic slide, (right) shuttle rod extensions for simple indexes, decimal, and ordinal values. Image reproduced courtesy The John Gabriel Byrne Computer Science Collection.

logarithmic index "slides" to implement his Irish Logarithms. Figure 4 (based on McQuillan's work[22]) shows a slide with annotations for the decimal, ordinal, and index numbers. For each ordinal value on the slide axis Y, there was a corresponding simple index value on the slide axis X. It can be seen that the sequence of the simple index values forms a monotonic logarithmic profile along the X-axis. It also shows a storage shuttle with a type rod representing the decimal value of "8" (represented by the ordinal value "3"). The principle of operation is that the slide moves up until it hits the rod. When it does so, it will have been displaced by a simple index of 3 units. By contrast for a shuttle with a type rod representing the decimal value of "0" (represented by the ordinal value "9"), the slide will be displaced by a simple index of 50 units. It is thought quite likely that ordinals were used in the storage shuttles as a proxy for the decimal operands in order that the series of slide profile changes in Y would be monotonic, as this would facilitate the progressive movement of the slides along the X axis for all possible decimal values.

The slide and shuttle Y-axis of Figure 4 will actually be edge-on, facing directly away from the viewer (reducing their visibility to lines as in Figure 3), but they are illustrated as shown to aid understanding. The outer and inner slides faced each other on a common "starting line" aligned with their simple index "0." To multiply, the slides were moved toward their respective shuttles. Figure 5 shows, for example, the leftmost slide when moved until it hits the outer shuttle left type rod representing decimal value "8," and therefore displaced by

simple index $Z_8 = 3$. Figure 5 also shows the inner slide when moved until it hits the inner shuttle type rod representing decimal value "9," and therefore displaced by simple index $Z_9 = 14$. Like a slide rule, the relative displacement of the two slides is then the compound index $Z_{72} = Z_{8*9} = Z_8 + Z_9 = 3 + 14 = 17$, representing the decimal partial product $8*9 = 72$. Thus the logarithm of the product of two digits is represented by the increase in overlap of the two slides in X after they have been displaced from their starting alignment.

The upper part of Figure 6 shows a six-digit outer shuttle and a four-digit inner shuttle, representing a machine with a six-digit outer operand (the multiplicand 813200) and a four-digit inner operand (the multiplier 9247) as per *Engineering*.[21] These correspond to the variable values to be multiplied, which are from the outer and inner storage cylinders, respectively. These inner and outer shuttles were in line as shown. Figure 6 also shows six outer slides and one inner slide. The two types of slide are likely to have had the same logarithmic profile, as is shown. Each of the outer slides was moved to hit a corresponding digit of the outer shuttle. Then, starting with the most significant digit, the inner slide moved one digit at a time to hit the less significant inner shuttle digits, while the Mill accumulated the succeeding partial products to produce the result of Ludgate's variant of long multiplication. Figure 6 shows the partial products "72," "9," "27," "18," "0," "0" for the six outer slides. This figure reproduces the example multiplication of the outer operand (813200, the multiplicand) by the most significant digit (9) of the inner operand (9247, the multiplier) of *Engineering*[21] as per Figure 3. Ludgate's 1909 paper includes an almost identical example.

Ludgate's index was more than just the slides and shuttles, and in fact was a quite complex multiplicity of moving parts. Much of this related to converting the relative displacement Z_{J*K} to discrete increments or decrements of a result held on the set of "figure wheels" that Ludgate called the Mill, in which digits were represented by the rotational positions of the wheels. The two slide types existed within different physical structures. There were 20 of the outer slides (with the pointer attached at one end) in an outer frame but free to move within it (so for illustrative purposes that frame can be treated as invisible). In contrast, there was just one inner slide, firmly attached to an inner frame containing multiple blades. There was one blade per compound index, i.e., one per uniquely valid partial product, as per *Engineering*[21] (see Figure 6). The relative position of the pointer over the blades indicated the partial product. The way in which the blades were composed and utilized was a very clever arrangement.

Investigating the Work and Life of Percy Ludgate

ARTICLE

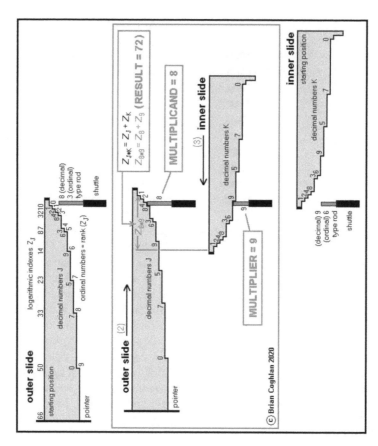

FIGURE 5. Multiplication using Ludgate's logarithmic slides for multiplication of 8∗9 = 72. Image reproduced courtesy The John Gabriel Byrne Computer Science Collection.

It is clear from Ludgate 1909 that the 20-digit multiply-accumulate procedure was as follows.

1) The outer (multiplicand **b**) and inner (multiplier **a**) operand shuttles are moved to the starting line "near the Index," with the inner slide aligned with the most-significant rod of the inner shuttle (the left rod).
2) The set of 20 outer slides move in one direction until they hit the rods of the shuttle holding the value of the outer operand. The slides convert the digits to "simple index numbers," essentially the logarithmic indexes of the digits of that operand.
3) Then, the single inner slide is moved in the opposite direction until it hits the rod of the shuttle that it is aligned with; the rod holds the value of that digit of the inner operand. The inner slide converts that digit to a simple index number, i.e., the logarithmic index of the value of that digit of that operand.
4) "As the index is attached to the last-mentioned slide, and partakes of its motion, the relative displacement of the index and each of the [outer

operand] slides … [and] pointers attached to the [outer operand] slides, which normally point to zero on the index, will now point respectively …" to the compound index number, essentially the sum of the simple index numbers, representing multiplication of the operands.
5) Next, all the compound index numbers are mapped by "movable blades" to numerical units and tens of the partial products and "conveyed by the pointer to" the Mill and accumulated in the Mill.
6) Then, the Index and its attached inner slide perform a "rapid reciprocating action" to align with the rod of the inner shuttle that holds the value of the next less significant digit of the inner operand.
7) Operations (3–6) are repeated for each of the rods of the inner shuttle "until the whole product of **ab** is found."
8) Then, "the shuttles are afterward replaced in the shuttle boxes."

These repetitions of operations for each of the rods of the inner operand digits until the whole

Coghlan et al, IEEE Annals of the History of Computing, Volume 43, Issue 1, pp. 19-37, 2021. © 2021, IEEE, reproduced with permission

Investigating the Work and Life of Percy Ludgate

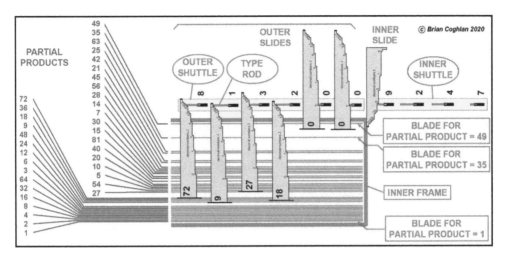

FIGURE 6. Ludgate's outer and inner logarithmic index slides, and inner frame with blades for compound indexes. Image reproduced courtesy The John Gabriel Byrne Computer Science Collection.

product is found can be observed in the cut-down six-digit by four-digit example in Figure 7. Comparison of the quadrants of Figure 7 very clearly shows how the outer slides do not move as the iterations of operations (3–6) above proceed. Only the inner slide and its frame move together in two dimensions (vertically up to retract, then horizontally right to the next digit, then vertically down to hit that digit's type rod) as they perform each "rapid reciprocating action" of operation 6) above to move digit-by-digit from the most to least significant digit of the inner operand.

Few details are available of how the partial products were accumulated. It is not possible in this short paper to go into the detail, but all this is considered in the (rather dense) analysis in Coghlan's work.[9] A surprising amount of understanding has emerged, conjectural construction detail has been uncovered, some useful dimensional, geometric and timing inequalities established, and an effort has been made to codify these new facts. Work has begun on both simulating parts of the machine and "reimagining" it with modern engineering software. Figure 8 shows early three-dimensional (3-D) renditions of the possible reimagined Ludgate races, shuttles, and Index performing the first iteration of the same multiplication as shown in Figure 7, i.e., positioned as shown in Figure 6 and in the top-left quadrant of Figure 7.

LUDGATE'S LIFE

In order to preserve this account's chronological flow the material below includes previously known facts from Randell, in each case clearly cited, e.g., as (Randell 1971[5]). Otherwise, the material below has all been discovered by the present investigation. Further extensive details of his ancestry collected by Brian Coghlan, Paul Hockie, and Trisha Gonzalez since 2016 are preserved for this account in our Collection[7] and online as Percy E. Ludgate: Parts A-C,[23–28] whereas updated material is aggregated in Percy E. Ludgate Prize.[8] These may be referred to for details of all births, marriages and deaths and other genealogical facts given in this paper, and much else. During the present investigation, archival records have been aggregated from a variety of official sources. Copies of these records are preserved in our Collection and also listed in Percy E. Ludgate: Part C, Evidence,[28] but permission will be required from those sources before these records can be made "open access," if ever.

Childhood

Percy Ludgate was born on August 2, 1883 in Townshend Street, Skibbereen (see Figure 9). We have traced the Ludgates back to his great-great-grandfather Matthew Ludgate, who lived at Marble Hill in the early 1700s.[25] His great-grandfather John Ludgate and grandfather Robert Ludgate farmed nearby.[29] His father was an ex-soldier, Michael Edward Ludgate, from Kilshannig, Mallow, Co. Cork. His mother Mary Ann Ludgate née McMahon, was born in Iden, Sussex, of an Anglo-Irish military family.[25,28] After Michael retired to Skibbereen in Co. Cork in 1876,[25] he in 1882 advertised as a shorthand tutor.[30] Percy was born in Skibbereen in 1883, the eighth and youngest child. From 1883 on there is a gap of seven years during which the family's whereabouts are as yet unknown.

From 1890 to 1898, Percy Ludgate's family then appear in Thom's Directory at 28 Foster Terrace,

ARTICLE

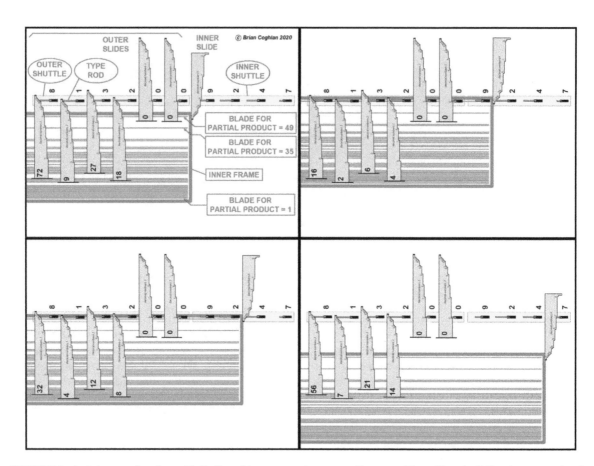

FIGURE 7. Ludgate's procedure for multiplication of two operands, representing a machine with a six-digit outer operand and a four-digit inner operand as per *Engineering*.[21] **Top Left**: iteration 1, multiplicand "813200" multiplied by multiplier digit "9" (from Figure 6). **Top Right**: iteration 2, multiplicand "813200" multiplied by multiplier digit "2." **Bottom Left**: iteration 3, multiplicand "813200" multiplied by multiplier digit "4." **Bottom Right**: iteration 4, multiplicand "813200" multiplied by multiplier digit "7." Between each quadrant of the figure, the Index and its attached inner slide move to the next multiplier digit. Note that the slides and shuttles will actually face directly away from the viewer, but are shown as is to aid understanding. Image reproduced courtesy The John Gabriel Byrne Computer Science Collection.

Dublin, as teachers of shorthand.[31] Clearly, this had become a small-scale family business. It was thought that 28 Foster Terrace no longer existed, but recently Gerry Kelly has discovered that the terrace was renumbered in the twentieth century, and that No. 28 is thought to have become No. 46 or No. 47,[32] both of which do still exist (see Figure 10).

From school records, it is known that during 1890–1891 Percy attended St. George's Infants School (aged 7–8).[24, p. 80] From 1892 to 1895, he is said to have attended St. George's National School (aged 8–12) (Randell 1971[5]), but there were two such schools in north Dublin, one behind St. George's Church in Temple Street, the other in North Strand, and for this period neither school's records have yet been found. His father was Episcopalian (a branch of Anglican Church),[33] and Percy attended St. George's Church (Randell 1971[5]).

Adolescence

From 1896 to 1898, Percy must have attended (an as yet unknown) secondary school, as it was compulsory until age 14,[34] and given his subsequent history must surely have excelled there.

In 1898, the *London Gazette* published that Percy Ludgate (then aged 15) was appointed a "Boy Copyist" in the Irish Civil Service[35]; this was a temporary post for boys aged 15–20 years. In 1899 (aged 16), his family moved to 30 Dargle Rd, Drumcondra, two miles northward,[36] see Figure 11, whereas his father Michael (giving an address in Balbriggan) was imprisoned in Kilmainham Gaol for nonpayment of debt.[25] In 1900

Coghlan et al, IEEE Annals of the History of Computing, Volume 43, Issue 1, pp. 19-37, 2021. © 2021, IEEE, reproduced with permission

Investigating the Work and Life of Percy Ludgate

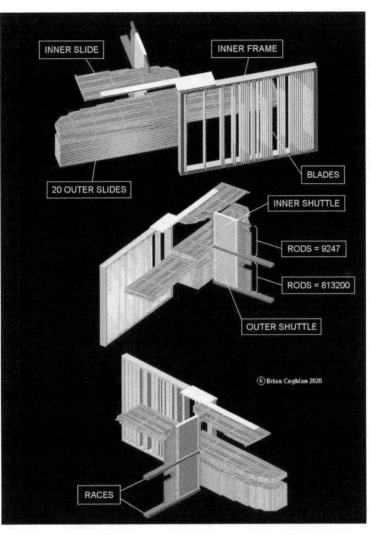

FIGURE 8. Early 3-D rendition of Ludgate's races, shuttles, and index for multiplication of two 20-digit operands, performing the first iteration of the same multiplication as shown in Figure 7, with multiplicand "813200" multiplied by multiplier "9247" as per *Engineering*[21] (the sign rod is at the top of each shuttle, next to the least-significant digit's type rod). **From top to bottom**: front, left-rear, and right-rear three-quarter views. Image reproduced courtesy The John Gabriel Byrne Computer Science Collection.

(when Percy was aged 17), his brother Frederick married Alice Walshe.[26, 28]

In 1901 (aged 18), the Irish Census lists Percy, his mother and a brother Alfred at 30 Dargle Rd. Percy is listed as a Civil Servant (Boy Copyist) in the National Education Office.[33] Frederick and Alice are listed at 24 Dargle Rd (just six doors away), and father Michael in Balbriggan (17 miles away). His other brother Thomas and wife Bedelia are listed in Cork.

In March 1903, Percy was the top Irish candidate in the Civil Service exams for Assistant Clerks.[37] Clearly, he must have excelled at Classics and Maths at school. He passed the medical, but for unknown reasons was not appointed to a position, whereas others

with lesser results were.[35] In August 1903 on his 20th birthday, Percy's Boy Copyist post expired, marking the end of his Civil Service career.[38] His daily occupation from 1903 is not yet known.

In that year he must have started work on his Analytical Machine, as his 1909 paper (which was submitted in December 1908) states it was the result of "six years' work." It is not obvious why he became interested in automating complicated mathematical calculations, whatever his exposure to standard calculating machines. Nor have we discovered obvious motivations or influences during our extensive ancestral searches. At this time very few calculating machines were based on multiplication, one exception being the

Investigating the Work and Life of Percy Ludgate

Coghlan et al, IEEE Annals of the History of Computing, Volume 43, Issue 1, pp. 19-37, 2021. © 2021, IEEE, reproduced with permission

FIGURE 9. 1798 Memorial, Skibbereen, Percy Ludgate's birthplace. Image reproduced courtesy of The Southern Star.

Millionaire calculator,[14] and only Thomas Fowler's ternary calculator c.1840[39] used rods or plates (as did Konrad Zuse's Z1 c.1937[40]). His niece said it was a private hobby, on which he did "work nightly, until small hours of the morning" (Randell 1971[5]). There remains the question of what he did in daylight hours.

Adulthood

In October 1904 (aged 21), Ludgate passed the Civil Service exams for Clerkships 2nd Division of the Civil Service (having passed 1st Division a year earlier), but failed the medical.[35] In February 1905 (aged 22), the "Case of Mr. Percy Ludgate – Irish Civil Service" was raised in the House of Commons, Westminster, by Timothy Harrington MP KC (who owned the United Ireland and Kerry Sentinel newspapers). The petition was for a new medical examination with a view to being certified for one or other of the positions for which he had

passed exams. He got a negative response from the Financial Secretary of the Treasury (soon to be Duke of Devonshire): "As nearly a year had elapsed since Mr. Ludgate's medical examination for an assistant clerkship, it was necessary to re-examine him before issuing him a certificate for a second-division clerkship. The result of the medical examination proving unsatisfactory the Civil Service Commissioners were unable to grant a certificate for either position."[41]

By the 1911 Census, Percy, his mother and Alfred were still at 30 Dargle Rd. Percy was listed as a commercial clerk to a corn merchant,[42] but it is not known when this began. It is possible it was a deliberate first step toward an accounting career.

In December 1908 (aged 25), he submitted his paper to the Royal Dublin Society, and in April 1909, "On a Proposed Analytical Machine" was published in their *Scientific Proceedings*.[1] In July 1909, a review by Prof. C. V. Boys of the paper was published in *Nature*.[2] Percy Ludgate's publication was a stunning achievement for a clerk to a corn merchant, to whom the RDS Science Committee, including several Fellows of the Royal Society and numerous professors, could well have been daunting. He and his proposed machine would have become known in what he may have thought an exalted circle.

In 1914 (aged 31), his paper "Automatic Calculating Machines"[43] was published in the *Handbook of the Napier Tercentenary Exhibition*, which took place in Edinburgh during July 24–27 (World War I started on July 28). This paper focused on Babbage, but mentioned Ludgate's 1909 design, and a subsequent difference engine design.[44] That Ludgate, rather than a notable mathematics professor, should describe Babbage's engines is curious.

FIGURE 10 . Two potential candidates for Percy Ludgate's first home in Dublin, images reproduced courtesy Gerry Kelly. (Left) 47 Foster Terrace, Ballybough, Dublin, and (right) 46 Foster Terrace, Ballybough, Dublin.

FIGURE 11. 30 Dargle Road, Drumcondra, Percy Ludgate's second home in Dublin. Image reproduced courtesy Michael Mongan.

FIGURE 12 . 31 Dame Street, Dublin, the location of Kevans & Son, where Percy Ludgate worked as an accountant. Image reproduced courtesy The John Gabriel Byrne Computer Science Collection.

During World War I (aged 31–35), he was engaged by a War Office committee (led by T.Condren-Flinn, of Kevans & Son accountants) on the production and sale of oats for cavalry, which involved planning and organization on a vast scale, and Ludgate was apparently praised for his contribution (Randell 1971[5]).

Another curiosity is that after the 1909 and 1914 papers there was very little mention of his work. Given C. V. Boys' 1909 report in *Nature*, and the c.1910 promotion by Henry Babbage of his father's machine, one might expect that the *Encyclopædia Britannica* would have mentioned those ideas, but the 1911 edition does not.[45] Maybe World War I upset the dissemination of Ludgate's ideas. Certainly, they had become relatively unknown by the time of Baxendall's well-known 1926 Science Museum catalogue of *Calculating Machines and Instruments*,[46] his mention in which stimulated Randell's 1971 investigation.

Maturity

It is commonly implied in relation to his 1909 paper that he was then an accountant, but that was not so until eight years later. For a period of at least two years,[47] Ludgate studied accountancy at the Rathmines College of Commerce (subsumed into the Dublin Institute of Technology in 1992[48]), then in 1917 (aged 34) he got a gold medal in Accountancy.[49] It is not known when he started working at Kevans & Son, 31 Dame Street (ultimately subsumed into Price Waterhouse Cooper, now known as PWC[50]) (see Figure 12). On any of his likely regular routes from home to work he could not have avoided seeing the impact of the Easter uprising of 1916 and the general mess created, much of which remained for years. Similarly, he could not have avoided witnessing the 1918-19 Spanish Flu epidemic, the War of Independence (1919–1921), or the start of the Civil War in June 1922.

FIGURE 13. Percy Ludgate, image reproduced courtesy Brian Randell.

He remained with Kevans & Son until 1922 as an accountant; a colleague stated he "possessed characteristics one usually associates with genius, . . . he was so regarded by his colleagues on the staff . . . humble, courteous, patient and popular." His niece Violet, daughter of Frederick and Alice, stated that "Percy . . . took long solitary walks . . . gentle, modest simple man," she ". . . never heard him make a condemning remark about anyone," she thought him ". . . a really good man, highly thought of by anyone who knew him," and that he ". . . Always appeared to be thinking deeply" (Randell 1971[5]). The only known photograph of him, provided by Violet to Randell, is shown in Figure 13, probably taken in the last five years of his life.

He never married, which is a pity as if he had we would probably have been able to find out more about his life.

Coghlan et al, IEEE Annals of the History of Computing, Volume 43, Issue 1, pp. 19-37, 2021. © 2021, IEEE, reproduced with permission

Investigating the Work and Life of Percy Ludgate

Investigating the Work and Life of Percy Ludgate

Coghlan et al, IEEE Annals of the History of Computing, Volume 43, Issue 1, pp. 19-37, 2021. © 2021, IEEE, reproduced with permission

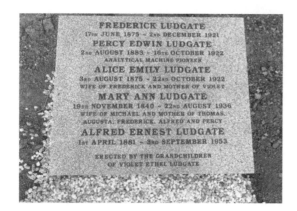

FIGURE 14. Ludgate grave after September 2019. Images reproduced courtesy The John Gabriel Byrne Computer Science Collection.

Tragic End

In December 1921, Ludgate's brother Frederick died of tuberculosis.[26,28] In October 1922, Ludgate developed pneumonia after a holiday in Lucerne, and on October 16, 1922, Percy Edwin Ludgate died aged 39, quickly followed on October 22, 1922 by the death of Frederick's wife Alice, she previously having nursed Percy (Randell 1971[5]).

In the UK and Ireland, life expectancy was 57 years in 1922,[51] and his family died at an average age of 58 years, so Percy died younger than expected. That Alice died six days later suggests a highly infectious illness. His death certificate states cause of death as catarrhal pneumonia (now bronchopneumonia), which mostly accompanies diseases like influenza.[52] The 1917–1920 Spanish Flu pandemic had a minor peak in 1922, e.g., in Bern in January 1922,[53] so it is possible Percy and then Alice contracted it.

Last Resting Place

Michael Ludgate died in 1923 in Belfast.[25] Percy's mother Mary died aged 96 in 1936. His remaining siblings Thomas died in 1951, Alfred in 1953, and Augusta in 1954, so by 1954 he, his parents and siblings had all died.[25,28] Those who had been living in Dublin were all buried in the same grave in Mount Jerome Cemetery.[54,24 pp. 95–99] His only living close relative, Violet, died in 1987. Percy's home from 1899 until his death in 1922, where he did his famous work, still exists in good condition.[24 pp. 27–29,108] On the other hand the Ludgate grave, occupied by Frederick, Percy, Alice, Mary and Alfred, and owned by Alfred,[55] who made no Will, lay unmarked and in poor condition in 2018, and could only be marked with permission from the owner, who was dead, or by close descendants, but there were no known descendants.[54] This was a very unsatisfactory situation given Percy Ludgate's importance to the history of computer science.

But then, astonishingly, we found a descendant of Violet! Subsequently, the following details emerged. In 1935, Violet gave birth to a daughter, who in 1936 was privately adopted and renamed, and was brought up overseas, married, gave birth to and raised six children. The children, who are direct descendants of three occupants of the Ludgate grave, were in 2019 allowed to erect a grave marker (see Figure 14). Thus, his ancestry,[25–27] which continues to be investigated as one possible path to finding preserved family material relating to Ludgate's life and work, has in fact first impacted progress on the previously intractable legal issue of marking the Ludgate grave.

Unknowns About Ludgate's Life

As details of his life are so scant, as of July 2020 there remain, despite our efforts, endless unknowns, for a sample see Table 5.

CONCLUDING REMARKS

Since Randell's 1971 paper on Ludgate numerous accounts of the history of computing have included

TABLE 5. Some known unknowns of Percy Ludgate's life.

Was he at North Strand National School?	Do Percy Ludgate's tax records exist?
Did he attend secondary school, and if so, where?	Do Prof. A.W.Conway's papers exist?
Which Corn Merchant did he join, and when?	Was he in poor health when he contracted pneumonia?
When did he start accountancy?	What ill health caused rejection by the Civil Service?
When did he join Kevans & Son?	Was there a relationship with Timothy Harrington MP?
Where is evidence of praise for his WW1 efforts?	Does his civil service file still exist in the UK?
Where are his extended relatives and acquaintances?	Do any family letters/photos/documents exist?

ARTICLE

APPENDIX

FROM: ENGINEERING, August 20, 1909, pp. 256–257[21]

Optical Character Recognition Courtesy
David McQuillan

A PROPOSED ANALYTICAL MACHINE

By name, at any rate, Babbage's famous analytical engine is known to all. It was intended to be a machine for the arithmetical solution of all problems in mathematical physics. Such solutions are generally, perhaps always, feasible, but in most cases when the computations have to be effected by direct human agency, they are so extremely tedious as to be practically, if not theoretically, impossible. Every operation in arithmetic can be reduced to addition, subtraction, multiplication, and division, and, indeed, the two latter operations can be regarded as mere extensions of the two former. The analytical engine was a machine by which these four operations could be performed in any desired sequence ; moreover, a number of partial operations could be combined, and the final results automatically tabulated for any required values of the variable. As is well known, though many years' labour was spent on the machine, it was never even partially completed, Mr. Babbage's scheme being far too ambitious for a first effort. He wished, indeed, to tabulate values to 50 significant figures, thus enormously complicating the mechanism and augmenting the cost of the experiment. In a paper read not long ago before the Royal Dublin Society, Mr. Percy E. Ludgate has revived again the idea of constructing such a machine. As proposed by him, the machine differs from that of Babbage in some fundamental details, though, as in its predecessor, Jacquard cards will be used to control the sequence of operations. Thus if, for instance, a number of values of the series

$$y = x - \frac{x^2}{2^2} + \frac{x^3}{2^2.3^2} - \frac{x^4}{2^2.3^2.4^2} + \&c.$$

were required, the appropriate card would be placed in the machine, which would then, for different values of *x*, calculate each term of the series, add all the positive terms together, subtract from this sum all the negative, and print the result. For a different series a different card would be used.

In Babbage's engine it was proposed to effect multiplication by successive additions, and divisions by successive subtractions, just as is now done in the case of the ordinary arithmometer. Mr. Ludgate, in his engine, proposes to effect these operations on entirely different principles. Multiplication is effected by a series of index numbers analogous to logarithms;

The arrangement is shown diagrammatically in Fig. 1 (see Figure 15). Here, the number 813,200 is to be multiplied by 9247. The arrow under 8 represents a slide, which to denote 8 is set at $\frac{3}{8}$ in. above the zero or starting line. The slide representing 1 lies on the starting line, whilst that representing 3 stands $\frac{7}{8}$ in. above this line, and that corresponding to the number two $\frac{1}{8}$ in. above. On the other hand, the slides representing zero are set 50 eighths above the starting line. The number of units above the starting line corresponding to each digit of the multiplicand are known as index numbers, and a complete table of these has been drawn up by Mr. Ludgate. All the slides aforementioned are mounted in a frame, and to multiply by 9, this frame is moved up over another frame divided with another series of index numbers. Thus, as shown, the distance between the lower frame and the starting line is such that the top of this lower frame lies on the index number corresponding to 9 ; that is, 14 eighths below the starting line. The lower end of the No. 8 slide, represented by the black circle, rests then, it will be seen, on a line marked "72," which is the product of 8 and 9. The digits 7 and 2 appear accordingly on the register below. Similarly, the tail of the No. 1 slide rests on the No. 9 line, that of the No. 3 slide on the No. 27 line, and that of the No. 2 slide on 18, corresponding to the partial products 9×1 ; 9×3 ; and 9×2. The tails of the zero slides rest on no line in the lower frame, and hence zero is registered for these. All these partial products are registered in the mill below, as indicated. In a final operation, these partial products are added together as indicated, giving 7,318,800. If now the frame is moved to the index number below the starting line marked "2," it will be found, on trial with a piece of tracing paper, that the tail of the No. 8 slide rests now on the line marked "16"—*i.e.*, 8×2. That of the No. 1 slide on the index-line marked "2," that of slide 3 on the line marked "15," and that of the No. 2 slide on the line marked "10." These partial products will then appear on the mill and be added together, giving the result of the multiplication of 813,200 by 2. The process is repeated for the remaining figures of the multiplier, and the whole added together so as to give the product of 813,200

Coghlan et al, IEEE Annals of the History of Computing, Volume 43, Issue 1, pp. 19-37, 2021. © 2021, IEEE, reproduced with permission

Investigating the Work and Life of Percy Ludgate

ARTICLE

× 9247. Mr. Ludgate proposes to give such products to twenty significant figures, the time required being, he states, about 10 seconds.

To divide one number by another he proceeds in a different fashion. He notes that the expression $\frac{p}{q}$, where p and q are any two numbers, can always be expressed in the form—

$$\frac{p}{q} = \frac{Ap}{1+x},$$

where x is a small quantity, and A is the reciprocal of some number between 100 and 999.

The above expression can also obviously be written

$$\frac{p}{q} = Ap\left(1 - x + x^2 - x^3 + x^4 - x^5, + \&c.\right),$$

the series being very rapidly convergent, the first eleven terms give the value of $\frac{1}{1+x}$ correct to at least twenty figures.

He proposes to perform division, therefore, by making the machine first calculate the value of this series, after which it will multiply A p by the value thus found. As a maximum, he considers that this operation giving the result correct to twenty figures might require $1\frac{1}{2}$ minutes.

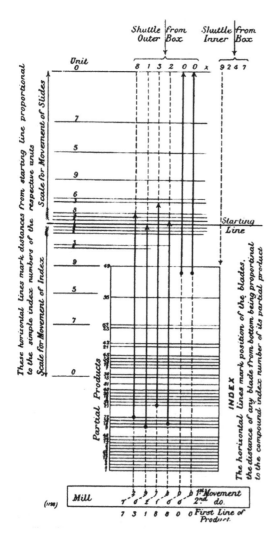

FIGURE 15. Diagram from *Engineering*, August 20, 1909, pp. 256–257.[21] Image reproduced courtesy Jade Ward of University of Leeds Library.

coverage of Ludgate's work, but no new material has been published. The new material in the present paper and considerable further material[8,9] has all been found since 2016, the majority since January 2017. By 2019 intensive genealogical efforts had identified his close family and their children, and his ancestors back to the early eighteenth century. His own life and early career, his close family members, domestic situation, legal issues, and 1909 paper submission, had all been explored but many issues remained outstanding.[24] In 2016, the Ludgate Hub was opened where Ludgate was born in Skibbereen.[56] With their help, in 2019 newspaper articles on Ludgate were published in *The Southern Star*[57] and *Cork Examiner* newspapers.[58] Subsequently an exploration into his life was presented to the August 2019 West Cork History Festival in Skibbereen,[59] and a public call for information was jointly announced by Trinity College Dublin and the Ludgate Hub. This led to our being contacted by Ralf Buelow at Christmas 2019, see below, plus a trickle of other responses. There is further aggregated information in the online catalog at Trinity College Dublin,[8] but clearly there are still many known unknowns relating to both Percy Ludgate's life and his work.

Our research has led to two significant discoveries, first at Christmas 2018 of the existence of Violet Ludgate's daughter and descendants, second at Christmas 2019 the discoveries of the articles and especially the diagram in *English Mechanic and World of Science*, then the discovery of the original of this material in *Engineering*, transcribed here as the Appendix. These led to further works, which are still in progress, first genealogical investigations (enabling the marking of the Ludgate grave),[60] and second efforts to deduce hidden details of Ludgate's analytical machine (the subject of the section titled "Reducing the Unknowns: New Discoveries" above regarding Ludgate's logarithmic multiplier, and more generally and extensively of Coghlan's work[9]).

ACKNOWLEDGMENTS

We thank Gerry Kelly (a pseudonymous contributor of detail on Ludgate's previously unknown civil service career), Prof. John Tucker, University College Swansea, U.K. (for access to the Riches thesis on Ludgate[61]), Canon Eithne Lynch, Mallow Church of Ireland (who has been very generous with her time looking up records), Dr. Susan Hood, RCB Library, Dublin (for help in accessing records), Royal Dublin Society Library (for access to records), Adrienne Harrington (Ludgate Hub), Victoria and Simon Kingston (West Cork History Festival), Lorcan Clancy (for audio and video recordings), and the Irish Government (for its generosity in establishing and populating their genealogy website http://www.irishgenealogy.ie/ with Irish civil and church records), the Gonzalez family (for their support), Ralf Buelow of Heinz Nixdorf MuseumsForum and Eric Hutton (for discovery of the articles in *English Mechanic and World of Science*), Jade Ward of the University of Leeds Library (for discovery of the article reproduced in the Appendix), and Eric Hutton, Jade Ward, and The Southern Star (for permission to publish the images). Finally, our thanks for the support of the School of Computer Science and Statistics, Trinity College Dublin, for this work and for The John Gabriel Byrne Computer Science Collection.[7]

REFERENCES/ENDNOTES

1. P. Ludgate, "On a proposed Analytical Machine," *Sci. Proc. Roy. Dublin Soc.*, vol. 12, no. 9, pp. 77–91, Apr. 28, 1909. Also reproduced in Ref. 5.
2. C. Boys, "A new Analytical Engine," *Nature*, vol. 81, pp. 14–15, Jul. 1909. Also reproduced in Ref. 5.
3. A. Lovelace, "Sketch of the Analytical Engine Invented by Charles Babbage, Esq. By L. F. Menabrea of Turin, Officer of the Military Engineers, With Notes by the Translator," in *Scientific Memoirs*, vol. 3, Ed., R. Taylor, London, U.K.: Richard and J.E. Taylor, 1843, pp. 666–731. [Online]. Available: https://www.scss.tcd.ie/SCSSTreasuresCatalog/literature/TCD-SCSS-V.20121208.870/TCD-SCSS-V.20121208.870.pdf
4. B. Collier, "Little engines that could've: The calculating machines of Charles Babbage," Ph.D. dissertation, Dept. History Sci., Harvard Univ., Cambridge, MA, USA, 1970. [Online]. Available: http://robroy.dyndns.info/collier/ Also available at Ref. 62 in file: "TheLittleEnginesThat-Couldve-TheCalculatingMachinesOfCharlesBabbage.pdf."
5. B. Randell, "Ludgate's Analytical Machine of 1909," *Computer J.*, vol. 14, no. 3, pp. 317–326, 1971. Available at Ref. 62 in file: "Randell-Ludgates-Analytical-Machine-of-1909-TheComputerJournal-1971-317-326.pdf."
6. B. Randell, "From Analytical Engine to electronic digital computer: The contributions of Ludgate, Torres, and Bush," *Ann. History Comput.*, vol. 4, no. 4, pp. 327–341, Oct. 1982. Available at Ref. 62 in file: "Randell-Contributions-of-Ludgate-Torres-and-Bush-IEEEAnnals-1982-4-4.pdf."
7. Trinity College Dublin, The John Gabriel Byrne Computer Science Collection. 2020. [Online]. Available: https://www.scss.tcd.ie/SCSSTreasures Catalog/
8. Trinity College Dublin, Percy E. Ludgate Prize in Computer Science. 2020. [Online]. Available: https://www.scss.tcd.ie/SCSSTreasuresCatalog/miscellany/TCD-SCSS-X.20121208.002/TCD-SCSS-X.20121208.002.pdf
9. B. Coghlan, Percy Ludgate's Analytical Machine, Trinity College Dublin. [Online]. Available: at Ref. 62 in file:"BrianCoghlan-PercyLudgatesAnalyticalMachine.pdf." 2020.
10. Royal Dublin Society, Minutes of the RDS Scientific Committee for 1908-1909, RDS Library, Dublin, Ireland.
11. B. Coghlan, personal communications with grandson of Prof. C. V. Boys, 2017-2018.
12. Irish Newspaper Archives, Communications, *Freemans Journal*, p. 6, Feb. 23, 1909, Also Irish Newspaper Archives, R. D. Society Scientific Meeting, Irish Independent, p. 6, Feb. 24, 1909.
13. Fujitsu, "FUJITSU Supercomputer PRIMEHPC FX1000: An HPC system opening up an AI and exascale era." 2020. [Online]. Available: https://www.fujitsu.com/downloads/SUPER/primehpc-fx1000-hard-en.pdf Also available at Ref. 62 in file: "Fujitsu-primehpc-fx1000-hard-en.pdf."
14. G. Saudan, *Swiss Calculating Machines, H. W. Egli A.-G. – A Success Story*, 147 p., self-published, Yens sur Morges, 2017.
15. W. McDonald, "J. Napier," *Dictionary of National Biography*, vol. 40. London, U.K.: Oxford Univ. Press, 1885–1900.
16. P. Hopp, *Slide Rules: Their History, Models and Makers*. Mendham, NJ, USA: Astragal Press, 1999.
17. E. Hutton, 2020. [Online]. Available: http://www.englishmechanic.com/. Also donation by Eric Hutton of digitized archives of English Mechanic and World of Science. [Online]. Available: https://www.scss.tcd.ie/SCSSTreasuresCatalog/literature/TCD-SCSS-V.20200520.001/TCD-SCSS-V.20200520.001.pdf

Coghlan et al, IEEE Annals of the History of Computing, Volume 43, Issue 1, pp. 19-37, 2021. © 2021, IEEE, reproduced with permission

Investigating the Work and Life of Percy Ludgate

18. R. Buelow, Percy Ludgate, der unbekannte Computerpionier {Percy Ludgate, the unknown computer pioneer}, Heinz Nixdorf MuseumsForum, Jan. 17, 2020, [Online]. Available: https://blog.hnf.de/percy-ludgate-der-unbekannte-computerpionier/. Also Google translation to English, courtesy Brian Randell. Available at Ref. 62 in file: "RalfBuelow-HNF-Blog20200117-translation-BRandell-v2-20200117-0756.pdf."

19. "Scientific News: No. 9 of the Scientific Proceedings of the Royal Dublin Society," *English Mechanic and World of Science*, no. 2302, p. 322, May 7, 1909, Discovered by Ralf Buelow, and image provided by Eric Hutton Dec. 19, 2019. Reproduced in Ref. 9, and also available at Ref. 62 in file: "THE-ENGINEER-image15657-RalfBuelow-EricHutton-20191220-0329.jpg."

20. "Engineering (attribution): A proposed analytical machine," *English Mechanic and World of Science*, vol. 90, no. 2319, p. 111, Sep. 3, 1909, discovered by Ralf Buelow, and image provided by Eric Hutton. Dec. 19, 2019. Reproduced in Ref. 9, and also available at Ref. 62 in file: "THE-ENGINEER-image14800-RalfBuelow-EricHutton-20191220-0329.jpg."

21. "A proposed Analytical Machine," *Engineering*, pp. 256–257, Aug. 20, 1909, discovered by Jade Ward, Leeds University Library, Jan. 14, 2020, OCR by David McQuillan, Feb. 3, 2020, and image provided by Jade Ward Jan. 14, 2020. Reproduced in Ref. 9, and also available at Ref. 62 in file: "ENGINEERING1909-JadeWard-UnivLeedsLib-2020114-1111.pdf."

22. D. McQuillan, "The feasibility of Ludgate's analytical machine." 2020. [Online]. Available: http://www.fano.co.uk/ludgate/Ludgate.html. Also available at Ref. 62 in file: "McQuillan-The-Feasibility-of-Ludgates-Analytical-Machine-20160126-0607.pdf."

23. Trinity College Dublin, Percy E. Ludgate: Part A1, Prize in Computer Science. Available at Ref. 62 in file:"TCD-SCSS-X.20121208.002-20191027-1533-Draft-partA1.pdf." 2020.

24. Trinity College Dublin, Percy E. Ludgate: Part A2, Extended Discussion. Available at Ref. 62 in file: "TCD-SCSS-X.20121208.002-20191027-1533-Draft-partA2.pdf." 2020.

25. Trinity College Dublin, "Percy E. Ludgate: Part B1a, genealogy report." Available at Ref. 62 in file: "TCD-SCSS-X.20121208.002-20191027-1533-Draft-partB1a-PercyLudgate-Report.pdf," and "Part B1b, Pedigree". Available at Ref. 62 in file: "TCD-SCSS-X.20121208.002-20191027-1533-Draft-partB1b-PercyLudgate-Pedigree-A3.pdf." 2020.

26. Trinity College Dublin, "Barbara Hopkins: Part B2a, genealogy report." Available at Ref. 62 in file: "TCD-SCSS-X.20121208.002-20191027-1533-Draft-partB2a-BarbaraHopkins-Report.pdf," and "Part B2b, Pedigree". Available at Ref. 62 in file: "TCD-SCSS-X.20121208.002-20191027-1533-Draft-partB2b-BarbaraHopkins-Pedigree-A3.pdf." 2020.

27. Trinity College Dublin, "Eileen Mary Ludgate: Part B3a, genealogy report." Available at Ref. 62 in file: "TCD-SCSS-X.20121208.002-20191027-1533-Draft-partB3a-EileenMaryLudgate-Report.pdf", and "Part B3b, Pedigree". Available at Ref. 62 in file: "TCD-SCSS-X.20121208.002-20191027-1533-Draft-partB3b-EileenMaryLudgate-Pedigree-A3.pdf." 2020.

28. Trinity College Dublin, Percy E. Ludgate: Part C, Evidence. Available at Ref. 62 in file: "TCD-SCSS-X.20121208.002-20191027-1533-Draft-partC.pdf." 2020.

29. Griffith Valuation, Parish of Kilshannig. 2020. [Online]. Available: http://www.askaboutireland.ie/griffith-valuation/

30. Irish Newspaper Archives, M. E. Ludgate, p. 3, *Skibbereen Eagle*, Sep. 16, 1882, Also: Irish Newspaper Archives, M. E. Ludgate, p. 4, *Skibbereen Eagle*, Nov. 18, 1882.

31. *Dublin Directory 1890–1898*, Thom's Directory, Dublin City Library and Archive, Gilbert Library, Dublin, 2020.

32. G. Kelly, 46 Foster Terrace, attached to personal email to Brian Coghlan, Nov. 9, 2019.

33. 1901 Census Return, 30 Dargle Road, Drumcondra, The National Archives of Ireland, 2020.

34. Hansard, Education (Ireland) Act 1892. [Online]. Available: https://api.parliament.uk/historic-hansard/acts/irish-education-act-1892 and https://en.wikipedia.org/wiki/Raising_of_school_leaving_age

35. *London Gazette*, pp. 3066–3068, May 22, 1896, pp. 3900–3901, Jul. 13, 1897, Jan. 14, 1898, pp. 6454–6455, Nov. 4, 1898, p. 6029, Sep. 19, 1902, p. 7095, Nov. 7, 1902, p. 1779, Mar. 17, 1903, pp. 5419–5420, Aug. 23, 1904.

36. *Dublin Directory 1899–1936*, Thom's Directory, Dublin City Library and Archive, Gilbert Library, Dublin, Ireland, 2020.

37. *Weekly Irish Times*, London Correspondence, Mar. 21, 1903.

38. G. Kelly, "Percy Edwin Ludgate—Irish civil service saga," attached to personal email to Prof. Brian Randell, Mar. 11, 2013.

39. M. Glusker, "Thomas Fowler's ternary calculating machine," *Brit. Soc. History Math.*, vol. 46, pp. 2–5, 2002. Also in: *J. Oughtred Soc.*, vol. 11, p. 2, 2002.

40. R. Rojas, "Konrad Zuse's legacy: The architecture of the Z1 and Z3," *Ann. History Comput.*, vol. 19, no. 2, pp. 5–15, 1997.

41. Hansard, vol. 141, Feb. 20, 1905. [Online]. Available: https://api.parliament.uk/historic-hansard/commons/1905/feb/20/irish-civil-service-case-of-mr-percy

42. 1911 Census Return, 30 Dargle Road, Drumcondra, The National Archives of Ireland, 2020.

43. P. Ludgate, "Automatic Calculating Machines," in: *Handbook of the Napier Tercentenary Celebration*, or *Modern Instruments and Methods of Calculation*, E. M. Horsburgh, Ed. Royal Society of Edinburgh, Edinburgh, and G. Bell and Sons Ltd, London, 1914.

44. B. Coghlan, "Speculations on Percy Ludgate's Difference Engine," Trinity College Dublin, in process. Available at Ref. 62 in file: "BrianCoghlan-Speculations-on-Percy-Ludgates-Difference-Engine.pdf." 2020.

45. Wikisource, 1911 Encyclopædia Britannica/Calculating Machines. [Online]. Available: https://en.wikisource.org/wiki/1911_Encyclop%C3%A6dia_Britannica/Calculating_Machines Also available at Ref. 62 in file: "Encyclopaedia-Britannica-11thEdition-Vol4Part4-1911-wikisource.pdf." 2020.

46. D. Baxandall, *Calculating Machines and Instruments: Catalogue of the Collection in the Science Museum*. London, U.K.: Science Museum, 1926.

47. *Irish Independent*, Irish Newspaper Archives, Corporate Accounts Exams, p. 4, Sep. 2, 1916.

48. Government of Ireland, Dublin Institute of Technology Act, 1992. [Online]. Available: http://www.irishstatutebook.ie/eli/1992/act/15/schedule/1/enacted/en/html Also available at Ref. 62 in file: "Dublin-Institute-of-Technology-Act-1992.pdf."

49. *Freeman's Journal*, Corporation of Accountants, Corporate Accountants, Results of the June Examination, p. 2, Sep. 15, 1917.

50. Obituary, Dan McGing, *Sunday Independent*, Independent News & Media PLC, Oct. 28, 2012.

51. M. Roser, "Life expectancy." [Online]. Available: https://ourworldindata.org/life-expectancy/ Also available at Ref. 62 in file: "Roser-Life-Expectancy-Our-World-in-Data-withIreland-20200929-1908.pdf." 2020.

52. G. McConnell, *A Manual of Pathology*. Philadelphia, PA, USA: Saunders, 1915.

53. K. Zürcher, M. Zwahlen, M. Ballif, H. Rieder, M. Egger, and L. Fenner, "Influenza pandemics and tuberculosis mortality in 1889 and 1918: Analysis of historical data from Switzerland," *PLoS One*, vol. 11, Oct. 5, 2016, Art. no. e0162575.

54. Mount Jerome Cemetery. 2020. [Online]. Available: http://www.mountjerome.ie/

55. P. Ludgate, NAI/CS/PO/TR Will of Percy Edwin Ludgate, National Archives of Ireland, Jun. 26, 1917.

56. N. Baker, "Potential €37m impact of digital life on Skibbereen," *Irish Examiner*, p. 7, Nov. 5, 2015.

57. D. Forsythe, "Could WW1 have ended Ludgate's computer dreams?" *The Southern Star*, Jul. 27, 2019.

58. K. O'Neill, "West Cork's own computer pioneer," *Cork Examiner*, Aug. 6, 2019.

59. B. Coghlan, "An exploration of the life of Percy Ludgate," West Cork History Festival, Skibbereen, Ireland, Aug. 10, 2019. Available at Ref. 62 in file: "BrianCoghlan-An-exploration-of-the-life-of-Percy-Ludgate-wAnim.pdf."

60. B. Coghlan, B. Randell, P. Hockie, T. Gonzalez, D. McQuillan, and R. O'Regan, "Percy Ludgate (1883-1922), Ireland's first computer designer," to be published.

61. D. Riches, "An analysis of Ludgate's Machine leading to the design of a digital logarithmic multiplier," Dept. Elect. Electron. Eng., Univ. College, Swansea, U.K., Jun. 1973. Available at Ref. 62 in file: "Riches-An-Analysis-of-Ludgates-Machine-UniversityCollegeSwansea-fromBrianRandell-20170203-1352.pdf."

62. Trinity College Dublin, Percy E. Ludgate Folder. 2020. [Online]. Available: https://www.scss.tcd.ie/SCSSTreasuresCatalog/miscellany/TCD-SCSS-X.20121208.002/. For a more mobile-friendly interface to selected Ludgate folder contents, see: https://www.scss.tcd.ie/SCSSTreasuresCatalog/ludgate/

BRIAN COGHLAN is a retired Senior Lecturer with the School of Computer Science and Statistics, Trinity College Dublin, The University of Dublin, Ireland. He is curator of The John Gabriel Byrne Computer Science Collection. Contact him at coghlan@cs.tcd.ie.

BRIAN RANDELL is an Emeritus Professor of Computing Science, and a Senior Research Investigator with the School of Computing, Newcastle University, U.K. His book *The Origins of Digital Computers: Selected Papers* was first published in 1973, by Springer Verlag. Contact him at Brian.Randell@ncl.ac.uk.

PAUL HOCKIE is a Genealogist from London, U.K. Contact him at paul@hockie.co.uk.

TRISH GONZALEZ is a Genealogist from Florida, USA. Contact her at trishalg03@aol.com.

DAVID MCQUILLAN is a retired Systems Designer from the U.K. who gives occasional talks on mathematics. Contact him at dmcq@fano.co.uk.

REDDY O'REGAN is a retired Solicitor from Skibbereen, Co. Cork, Ireland. Contact him at oreganic@live.ie.

Coghlan et al, IEEE Annals of the History of Computing, Volume 43, Issue 1, pp. 19-37, 2021. © 2021, IEEE, reproduced with permission

Investigating the Work and Life of Percy Ludgate

Investigating the Work and Life of Percy Ludgate

Coghlan et al, IEEE Annals of the History of Computing, Volume 43, Issue 1, pp. 19-37, 2021. © 2021, IEEE, reproduced with permission

1

Percy Ludgate's Analytical Machine

Brian Coghlan

Abstract This document attempts to deduce additional detail about Percy Ludgate's Analytical Machine by contextual analysis of both his 1909 paper and recently discovered articles. The aim is to assemble a body of known new design facts. Thus far, this has been found possible in relation to his logarithmic indexes, data format, Index and Mill, carry propagation, timing, division algorithm, storage, instruction set and preemption. The approach has proven to be surprisingly fruitful, and has allowed some progress to be made towards a 3D-printed re-imagining of the machine. This paper is a work in progress.

Index Terms— **Percy Ludgate, Analytical Machine, Irish Logarithms, Multiply-Accumulate (MAC), Division by Convergent Series**

I. INTRODUCTION

Historically, computing is generally viewed as having four fore-fathers: Charles Babbage (1843: the concept an analytical engine [4]), George Boole (1854: Boolean logic [24]), Alan Turing (1936: theory of computing [25]), Claude Shannon (1937: digital logic [26] and 1948: information theory [27]). Boole, Turing & Shannon impact on all aspects of modern society (principally because computing now does), Babbage does not. Nonetheless Babbage is very notable for introducing the concept of the first analytical engine in history. His importance is historical. He also raised the idea of thinking machines, a controversial subject then, like AI is today.

Babbage's "Analytical Engine" [53] was a very novel concept, with processing done in a Mill based on addition, where the Mill and Storage were made of clockwork cogs and gears (Liebniz c.1671 [13]). Programming and input/output were done via punched cards (Jacquard c.1804 [14]). Ada Lovelace published Babbage's design in 1843 [4]. Only a small portion of it was ever built, but Babbage left very extensive 2-d drawings (for 3-d graphic art representations, see [12]), and these are now being put into modern engineering software [15][18], so perhaps more of it will yet be built.

Percy Edwin Ludgate (1883-1922) is notable as the second person to publish a design for his "Analytical Machine" [1][2][16], the first after Lovelace published Babbage's design. Ludgate's machine included a Mill, but also an "Index" to do multiplication. The combination did multiply-accumulation (MAC), which is now important in signal processing [19], e.g. in radar and astronomy, and in deep AI [20]. The multiply embodied a discrete table, as did a few other calculating machines at this time, such as the Millionaire [21]. But it was based on a novel concept now called "Irish Logarithms", and worked analogously to a slide rule (Oughtred c.1622 [23]). The accumulation worked like Napier's "bones" (Napier c.1614 [22]), otherwise called "rods", once widely used to facilitate the accumulation of partial product units and tens digits inscribed on their upper surface.

These were the only two mechanical designs well before the electronic computer era. Subsequently, c.1914 electromechanical designs began, and c.1937 electronic designs began [11]. A third mechanical design arose at the dawn of the electronic computer era: in 1936 Zuse's "V1", later renamed "Z1". Historically there are four types of Analytical Machines: Babbage (mechanical, novel, 1843), Ludgate (mechanical, very different, 1909), Torres y Quevedo and his successors (electromechanical, 1914 onwards [7]), then modern computers and their billions of successors (fully electronic, 1949 onwards). Analytical machines now pervade, running big data, big science, health, business, government, and now domestic appliances, mobile phones, and even toys. Ludgate is notable for the second published analytical machine design in history, and Irish Logarithmic indexes, and the first multiply-accumulator (MAC) in a computer, and the first division by convergent series in a computer, and very novel concepts for storage and programming (e.g. two-operand instructions, pre-emption). Like Babbage, his importance is historical.

Ludgate and Babbage appear to have had no influence on modern computers, since as far as is known there is no record in the literature of modern electronic computers that specifically states inheritance from either of their machines. Randell's 1971 paper [6] was followed by just one attempt, in 1973, by Riches, at a conjectural machine design based on Ludgate's design [42]. Copies of most of the related literature are held by The John Gabriel Byrne Computer Science Collection [8] (and its host department presents an annual Ludgate Prize [9]). This and [10] outline the initial technical results of an investigation by the Collection (and [17] outlines its initial historical results). The extremely few other technically-related efforts are referred to further below.

Dr.Brian Coghlan is with the School of Computer Science and Statistics, Trinity College Dublin, The University of Dublin, Ireland (e-mail: coghlan@cs.tcd.ie).

BrianCoghlan-PercyLudgatesAnalyticalMachine-20220320-1305.doc

2

II. ASPECTS OF LUDGATE'S MACHINE

II.1. Features

Only a few features are described in Ludgate's 1909 paper. The rest must be deduced by contextual analysis of the paper, largely a process of logical inference or elimination of false propositions, and mechanical or mathematical speculations. They are summarised in Table 1, but explored further below.

	Base operation is multiply-accumulate (MAC) not addition
	Multiply is done with Irish Logarithms by an INDEX
	Long multiply starts at left digit of multiplier
¥	Numbers must be fixed-point
	Multiply-accumulate partial-products are added units first, then tens by a MILL
¥	Timing implies pipelining tens carry adds
¥	Instruction set: ADD, SUBTRACT, MULTIPLY, DIVIDE, LOG, STORE, CONDITIONAL BRANCH
	Two-operand addressing for load
¥	Two-operand addressing for store
	Fast for 1909: ADD / SUB 3 sec, MUL 10 sec, DIV 90 sec, LOG 120 sec
¥	Storage of 192 variables implies (64 inner + 128 outer) shuttles, equispaced
¥	Hence storage size implies binary storage addressing
	Numbers stored via rod for sign & every digit protruding 1–10 units
	Data input / output via punched number-paper (or upper keyboard)
	Program input / output via punched formula-paper (or lower keyboard), one instruction per row
¥	Manual preemption
¥	Measurements in units of an eighth of an inch
¥	Main shaft gives 'cycle time' of a third of a second
	Small size: estimated by Ludgate as 0.5m H x 0.7m L x 0.6m W

Table 1 *Features of Percy Ludgate's analytical engine, with inferences denoted by ¥*

II.2. Unknowns

Almost everything about its construction is unknown, for a small sample of these unknowns, see Table 2. Many such issues are explored further below.

How shuttles were selected in storage cylinders	Any internal dimensions
How a shuttle was moved	Any internal timing
How the INDEX mechanism worked	Almost everything about program control
How the MILL mechanism worked	Almost everything about input & output

Table 2 *Sample of known unknowns of Percy Ludgate's analytical engine*

II.3. Reducing the Unknowns: New Discoveries

Clearly the discovery of drawings by Ludgate would reduce the unknowns, so this has been an ultimate target of the research, but any discovery of new information about Ludgate's design would help. And in fact, just before Christmas 2019, Ralf Beulow of Heinz Nixdorf MuseumsForum with Eric Hutton [28] discovered a pair of articles about Ludgate's machine in the 1909 issues of the popular science magazine "The English Mechanic and World of Science" [29]. One was a very brief summary [30] (see Appendix II) of Ludgate's 1909 paper, with no new information.

The other article [31] (see Appendix III) did indeed promise new information, if validated, and included the first known diagram of Ludgate's machine! This article was attributed to "Engineering". A search by Jade Ward of University of Leeds Library discovered the article was in fact an abbreviated extract of an article [32] published a month earlier, i.e. just over three months after Ludgate's original Royal Dublin Society paper, in "Engineering" (a London-based monthly magazine founded in 1865). This article, which from here onwards will be referred to as "Engineering [32]", is reproduced in full as Appendix I.

3

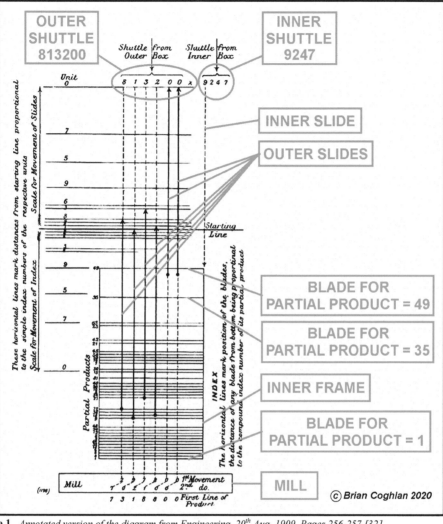

Figure 1 *Annotated version of the diagram from Engineering, 20th Aug. 1909, Pages 256-257 [32]*
See Appendix I for the original full-size image
Image reproduced courtesy The John Gabriel Byrne Computer Science Collection

Figure 1 shows an annotated version of the diagram from Engineering [32] (an identical copy of which appeared in "The English Mechanic and the World of Science" [31]), which presumably was provided, and perhaps actually drawn, by Ludgate. It is not yet known whether its text is handwritten or typeset. The *'1198'* near the bottom left of the diagram could be the publisher's notation, or could be a datestamp ('11-Sep-1908' encoded as *'1198'*), or could be Ludgate's sheet number (large but perhaps not impossible after years of redesigning). It would also be useful to have the text professionally analysed against excerpts from Percy Ludgate's 1917 Will and Testament [33].

The text of the article of Engineering [32] at first appeared to us to contradict Ludgate's 1909 paper, but as analysis has progressed it has begun to seem more likely that the text was provided by Ludgate too (the article includes a detailed explanation of the diagram). Ludgate and the writer of the article are quite precise in what they say. In fact treating Ludgate's paper and the article as equally valid has uncovered ways the design may have been that would have been very hard to arrive at with only Ludgate's 1909 paper. A surprising amount of understanding has emerged, construction detail has been uncovered, some useful dimensional, geometric and timing inequalities established, and an effort has been made to codify these new facts. All this is described further below. Work has begun on both simulating the machine and "re-imagining" it with modern engineering software.

4

III. ESSENTIAL PRINCIPLES OF LUDGATE'S MACHINE

III.1.Ludgate's multiply-accumulate operation

The multiplication in Ludgate's multiply-accumulate (MAC) operation is done in reverse to its traditional order. This is best illustrated graphically. For comparison, the traditional long multiplication ordering, starting with the multiplier least significant digit, is first illustrated in Figure 2.

TRADITIONAL	STEPS			TRADITIONAL	STEPS			TRADITIONAL	STEPS		
314	digit			314	digit			314	digit		
x 679	3			x 679	2			x 679	1		
------				------				------			
2826	<--	314 x	9	2826	<--	314 x	9	2826	<--	314 x	9
				2198	<--	314 x	70	2198	<--	314 x	70
								1884	<--	314 x	600
------				------				------			
282				24806				213206			© Brian Coghlan 2020

Figure 2 *Traditional long multiplication: multiply by LS digit of multiplier to form LS partial product, then multiply by next left digit of multiplier, and etc. For each step accumulate partial result to yield a final result (where LS and MS denote least and most significant)*
Image reproduced courtesy The John Gabriel Byrne Computer Science Collection

Ludgate's multiply-accumulate is processed jointly in his Index and Mill. The Index multiplies based on logarithms, while the Mill adds via clockwork cogs and gears (Ludgate refers to these as "wheels"). The multiply is done iteratively per digit from the most significant (MS) digit of the multiplier towards the least significant (LS) digit, i.e. in the reverse order to that in Figure 2. Figure 3 shows the first iteration (for the MS digit). For each digit of the multiplier, the units of the partial product are generated and added to the accumulator. Then the tens of the partial product are generated and added to the accumulator.

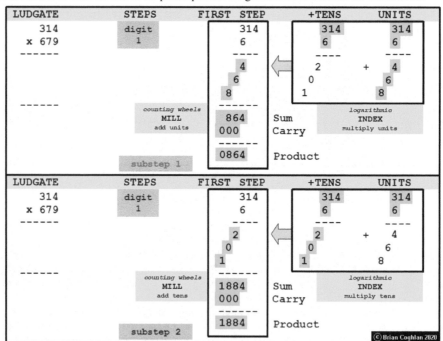

Figure 3 *Ludgate's long multiplication steps for the multiplier MS digit. First multiply-accumulate units, then tens*
Image reproduced courtesy The John Gabriel Byrne Computer Science Collection

Ludgate's 1909 paper stated "A carriage near the index now moves one step to effect multiplication by 10", and "After this the index makes a rapid reciprocating movement". This implies an awkward step left for tens then two steps right for the units of the next multiplier digit. This appears inexplicable, as it seems to be easier to add tens first then units for a smooth traverse from left to right, however it may have been an unknown optimization.

5

He also stated "I have devised a method in which the carrying is practically in complete mechanical independence of the adding process, so that the two movements proceed simultaneously". Indeed his stated operation timings imply that carries are added in parallel with subsequent mechanical movements until a final visible addition of carries after the last of the multiplication steps. The implications are considered later; for now, let us ignore these details.

Multiplication for the remaining digits is illustrated in Figure 4.

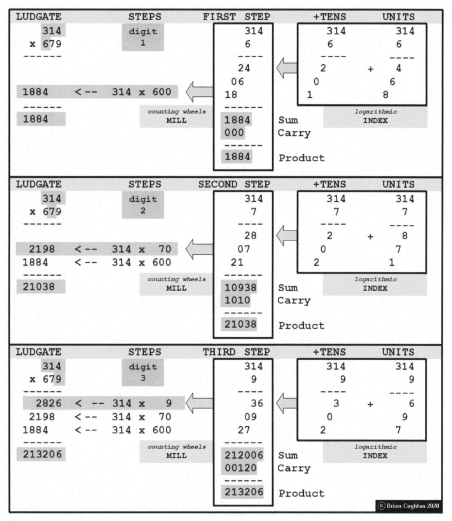

Figure 4 *A possible order in which Ludgate's machine would have calculated 314×678 with one add and carries per stage.*
The right hand side shows how the partial products are split into tens and units
Image reproduced courtesy The John Gabriel Byrne Computer Science Collection

Figure 5 shows a more succinct illustration of another example, but excluding any treatment of Ludgate's carry propagation, which is explored much further below. This example shows the calculation in Ludgate 1909, illustrating again the novel iteration order and how the partial products are split into units and tens, where the units of the partial product are generated by the Index and added to the accumulator by the Mill, then the tens of the partial product are generated and added to the accumulator.

Percy Ludgate's Analytical Machine

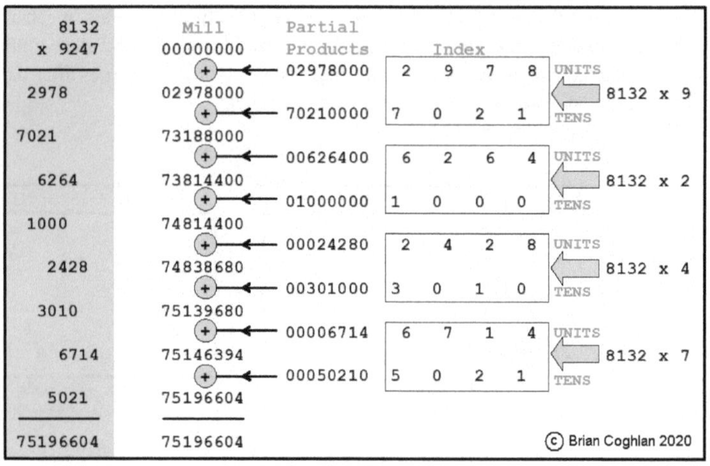

Figure 5 *An example of Ludgate's variant of long multiplication, calculating 8132 x 9247*
Image reproduced courtesy The John Gabriel Byrne Computer Science Collection

III.2. Ludgate's logarithmic indexes

Ludgate invented his own logarithmic indexes for multiplication, an entirely new result in 1909. As indicated above, these are not ordinary logarithms, but integer values that obey the logarithmic law. For two operands Z_J and Z_K, Ludgate's index numbers ensure $Z_Y = Z_{J*K} = Z_J + Z_K$. For example, indexes $Z_3 = 7$ and $Z_5 = 23$, therefore $Z_{15} = Z_{3*5} = Z_3 + Z_5 = 30$. As can be seen in Table 3, simple indexes form a non-monotonic function of the decimal operands, but a monotonic function of the ordinals (because ordinals represent the rank order of simple indexes). Therefore ordinals may be used as a proxy for the decimal operands in circumstances where monotonicity is desirable, and Ludgate employed this.

Decimal operand	Simple index	Ordinal number	Partial product	Compound index	Partial product	Compound index	Partial product	Compound index
0	50	9	1	0	15	30	36	16
1	0	0	2	1	16	4	40	26
2	1	1	3	7	18	15	42	41
3	7	4	4	2	20	25	45	37
4	2	2	5	23	21	40	48	11
5	23	7	6	8	24	10	49	66
6	8	5	7	33	25	46	54	22
7	33	8	8	3	27	21	56	36
8	3	3	9	14	28	35	63	47
9	14	6	10	24	30	31	64	6
			12	9	32	5	72	17
			14	34	35	56	81	28

Table 3 *Ludgate's simple and compound logarithmic indexes (reproduced from Ludgate's 1909 paper, Table 1 and 2)*

C.V.Boys said: "Ludgate … uses for each of the prime numbers below ten in a logarithmic system with a different incommensurable base, which as a fact never appears" [2], i.e. each $Z_X = \log_{N_X}(X)$ has a different (invisible) base N_X.

Logarithmic indexes were not new in 1909, as Jacobi/Zech indexes (which can be derived from number theory) were already in use in astronomy, but Ludgate's particular indexes were new. Ludgate said: "The index numbers (which I believe to be the smallest whole numbers that will give the required results)". The largest Z_{J*K} is $Z_{7*7} = 66$. By way of comparison:

(1) Jacobi/Zech indexes (1846/1849 [35]). For example $Z_1 = 0$, $Z_2 = 1$, $Z_3 = 18$, $Z_5 = 44$, $Z_7 = 7$.
An alternative is $Z_1 = 0$, $Z_2 = 1$, $Z_3 = 8$, $Z_5 = 44$, $Z_7 = 27$. In both cases the largest Z_{J*K} is $Z_{5*5} = 88$.
(2) Remak indexes (Von K. Hoecken 1913 [36]), $Z_1 = 0$, $Z_2 = 1$, $Z_3 = 13$, $Z_5 = 21$, $Z_7 = 30$, where the largest Z_{J*K} is $Z_{7*7} = 60$.
(3) Korn indexes (Von K. Hoecken 1913 [36]), $Z_1 = 0$, $Z_2 = 8$, $Z_3 = 13$, $Z_5 = 1$, $Z_7 = 30$, where the largest Z_{J*K} is $Z_{7*7} = 60$.

Ludgate's focus on small numbers makes it possible he knew of Jacobi/Zech indexes. Remak and Korn indexes were reported after Ludgate's 1909 paper was published. Small numbers are an important metric used to minimize the length of any mechanism that is used in the machine design to convert to or from the logarithmic indexes. McQuillan [37][38] has shown that better indexes can readily be computed using modern machinery.

Lemma 1: Ludgate proposed logarithmic indexes with the smallest Z_{J*K} then known.

As happens surprisingly often with novel inventions, in the same year as Ludgate's 1909 paper, a slide rule with Jacobi indexes was designed by Prof.Schumacher of Germany [39] and later manufactured as the Faber Model 366 [40]. For an example of how logarithmic indexes are used in calculations, see Andries de Man's educational emulator [41] for Ludgate's indexes.

III.3. Algorithm to derive Irish Logarithms

Only Boys [2], Riches [42] and de Man [34] offer any guidance as to how Ludgate derived some or all of his indexes. He almost certainly did not derive his indexes from number theory, but developed a systematic "method", such as that in Figure 6. The method of Figure 6 can be expressed as a simple algorithm, for example in Python as in Figure 7. Executing this algorithm yields identical indexes to those given in Tables 1, 2 and 3 of Ludgate's 1909 paper, see Figure 8.

Lemma 2: Ludgate's proposed logarithmic indexes are amenable to construction by a systematic method.
Lemma 3: Ludgate's proposed logarithmic indexes are amenable to algorithmic construction.

Irish Logarithms might exist for the range up to any N, not just for 0-9 as in Ludgate's paper. Certainly they have been shown to exist for 0-99 by extending the rather inefficient algorithm of Figure 7 with a basic Sieve of Eratosthenes to generate the seed prime numbers. Beyond that a more efficient algorithm would be necessary, or ideally a number-theoretic proof.

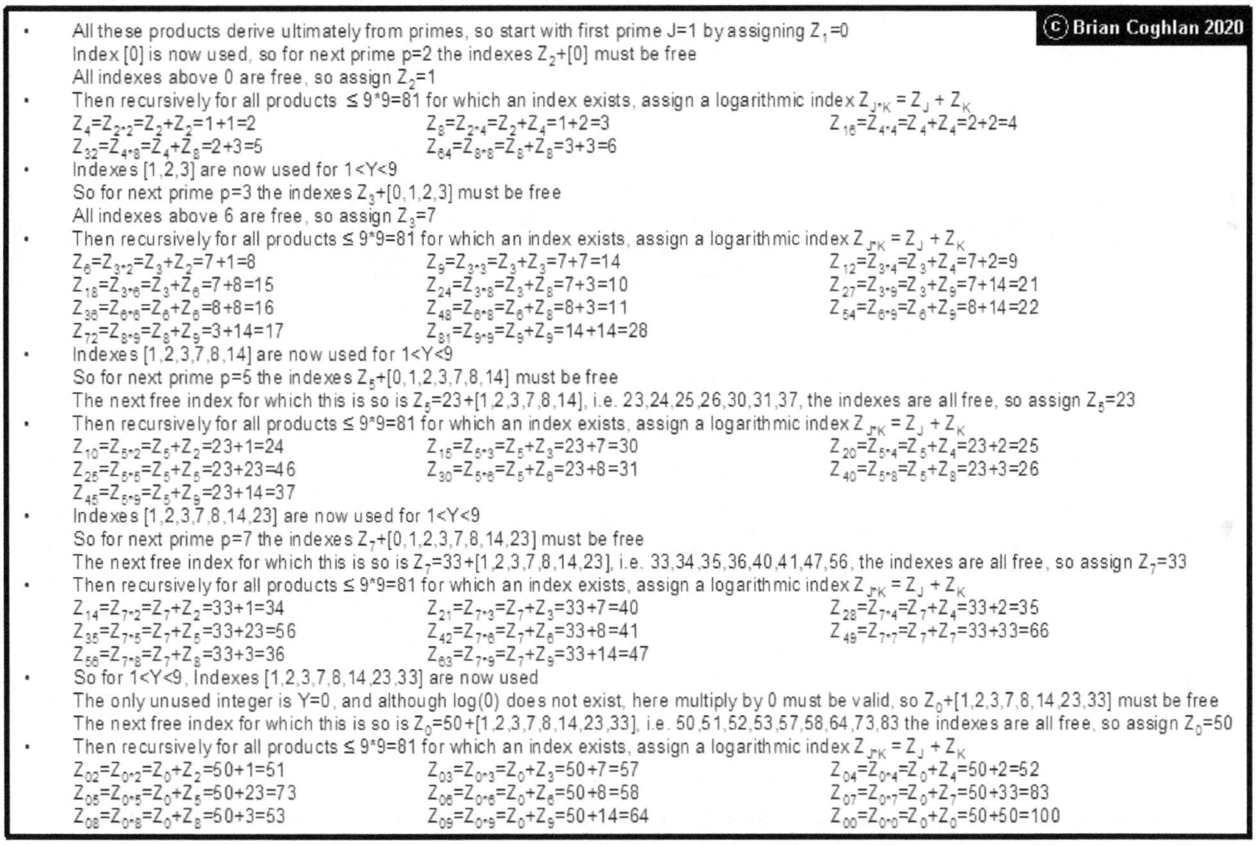

- All these products derive ultimately from primes, so start with first prime J=1 by assigning $Z_1=0$
  Index [0] is now used, so for next prime p=2 the indexes $Z_2+[0]$ must be free
  All indexes above 0 are free, so assign $Z_2=1$
- Then recursively for all products ≤ 9*9=81 for which an index exists, assign a logarithmic index $Z_{J*K} = Z_J + Z_K$
  $Z_4=Z_{2*2}=Z_2+Z_2=1+1=2$ $Z_8=Z_{2*4}=Z_2+Z_4=1+2=3$ $Z_{16}=Z_{4*4}=Z_4+Z_4=2+2=4$
  $Z_{32}=Z_{4*8}=Z_4+Z_8=2+3=5$ $Z_{64}=Z_{8*8}=Z_8+Z_8=3+3=6$
- Indexes [1,2,3] are now used for 1<Y<9
  So for next prime p=3 the indexes $Z_3+[0,1,2,3]$ must be free
  All indexes above 6 are free, so assign $Z_3=7$
- Then recursively for all products ≤ 9*9=81 for which an index exists, assign a logarithmic index $Z_{J*K} = Z_J + Z_K$
  $Z_6=Z_{3*2}=Z_3+Z_2=7+1=8$ $Z_9=Z_{3*3}=Z_3+Z_3=7+7=14$ $Z_{12}=Z_{3*4}=Z_3+Z_4=7+2=9$
  $Z_{18}=Z_{3*6}=Z_3+Z_6=7+8=15$ $Z_{24}=Z_{3*8}=Z_3+Z_8=7+3=10$ $Z_{27}=Z_{3*9}=Z_3+Z_9=7+14=21$
  $Z_{36}=Z_{6*6}=Z_6+Z_6=8+8=16$ $Z_{48}=Z_{6*8}=Z_6+Z_8=8+3=11$ $Z_{54}=Z_{6*9}=Z_6+Z_9=8+14=22$
  $Z_{72}=Z_{8*9}=Z_8+Z_9=3+14=17$ $Z_{81}=Z_{9*9}=Z_9+Z_9=14+14=28$
- Indexes [1,2,3,7,8,14] are now used for 1<Y<9
  So for next prime p=5 the indexes $Z_5+[0,1,2,3,7,8,14]$ must be free
  The next free index for which this is so is $Z_5=23+[1,2,3,7,8,14]$, i.e. 23,24,25,26,30,31,37, the indexes are all free, so assign $Z_5=23$
- Then recursively for all products ≤ 9*9=81 for which an index exists, assign a logarithmic index $Z_{J*K} = Z_J + Z_K$
  $Z_{10}=Z_{5*2}=Z_5+Z_2=23+1=24$ $Z_{15}=Z_{5*3}=Z_5+Z_3=23+7=30$ $Z_{20}=Z_{5*4}=Z_5+Z_4=23+2=25$
  $Z_{25}=Z_{5*5}=Z_5+Z_5=23+23=46$ $Z_{30}=Z_{5*6}=Z_5+Z_6=23+8=31$ $Z_{40}=Z_{5*8}=Z_5+Z_8=23+3=26$
  $Z_{45}=Z_{5*9}=Z_5+Z_9=23+14=37$
- Indexes [1,2,3,7,8,14,23] are now used for 1<Y<9
  So for next prime p=7 the indexes $Z_7+[0,1,2,3,7,8,14,23]$ must be free
  The next free index for which this is so is $Z_7=33+[1,2,3,7,8,14,23]$, i.e. 33,34,35,36,40,41,47,56, the indexes are all free, so assign $Z_7=33$
- Then recursively for all products ≤ 9*9=81 for which an index exists, assign a logarithmic index $Z_{J*K} = Z_J + Z_K$
  $Z_{14}=Z_{7*2}=Z_7+Z_2=33+1=34$ $Z_{21}=Z_{7*3}=Z_7+Z_3=33+7=40$ $Z_{28}=Z_{7*4}=Z_7+Z_4=33+2=35$
  $Z_{35}=Z_{7*5}=Z_7+Z_5=33+23=56$ $Z_{42}=Z_{7*6}=Z_7+Z_6=33+8=41$ $Z_{49}=Z_{7*7}=Z_7+Z_7=33+33=66$
  $Z_{56}=Z_{7*8}=Z_7+Z_8=33+3=36$ $Z_{63}=Z_{7*9}=Z_7+Z_9=33+14=47$
- So for 1<Y<9, Indexes [1,2,3,7,8,14,23,33] are now used
  The only unused integer is Y=0, and although log(0) does not exist, here multiply by 0 must be valid, so $Z_0+[1,2,3,7,8,14,23,33]$ must be free
  The next free index for which this is so is $Z_0=50+[1,2,3,7,8,14,23,33]$, i.e. 50,51,52,53,57,58,64,73,83 the indexes are all free, so assign $Z_0=50$
- Then recursively for all products ≤ 9*9=81 for which an index exists, assign a logarithmic index $Z_{J*K} = Z_J + Z_K$
  $Z_{02}=Z_{0*2}=Z_0+Z_2=50+1=51$ $Z_{03}=Z_{0*3}=Z_0+Z_3=50+7=57$ $Z_{04}=Z_{0*4}=Z_0+Z_4=50+2=52$
  $Z_{05}=Z_{0*5}=Z_0+Z_5=50+23=73$ $Z_{06}=Z_{0*6}=Z_0+Z_6=50+8=58$ $Z_{07}=Z_{0*7}=Z_0+Z_7=50+33=83$
  $Z_{08}=Z_{0*8}=Z_0+Z_8=50+3=53$ $Z_{09}=Z_{0*9}=Z_0+Z_9=50+14=64$ $Z_{00}=Z_{0*0}=Z_0+Z_0=50+50=100$

© Brian Coghlan 2020

Figure 6 *Systematic method to derive Ludgate's index numbers*
For all products $Y = (1<J<9) * (1<K<9)$ assign logarithmic index numbers $Z_Y = Z_{J*K} = Z_J + Z_K$
Image reproduced courtesy The John Gabriel Byrne Computer Science Collection

```
#!/usr/bin/env python                                    © Brian Coghlan 2020

import sys

# initialise variables
Z=[-1]*200    # table of complex indexes
PP=[-1]*200   # table of partial products
i = 0;
for p in (1,2,3,5,7,0):
    if Z[p]==-1:  # prime not indexed yet
        free=False
        while free==False and i<=100:
            free=True
            for j in (1,2,3,4,5,6,7,8,9):
                if free==True:
                    if Z[j]<>-1:                    # for existing indexes
                        for k in range (1,100):
                            if Z[k]==(i+Z[j]): # check if complex index exists
                                free=False
                                i=i+1
        if free==True: # OK, found a desired free set of indexes
            Z[p]=i        # create new simple index
            PP[i]=p       # create new partial product
            i=i+1
            for j in (1,2,3,4,5,6,7,8,9,0):
                if Z[j]<>-1:  # multiplicand simple index exists
                    for k in (1,2,3,4,5,6,7,8,9,0):
                        if Z[k]<>-1:  # multiplier simple index exists
                            if PP[Z[j]+Z[k]]==-1:     # product not indexed yet
                                if Z[j*k]==-1:
                                    Z[j*k]=Z[j]+Z[k] # create new complex index
                                    PP[Z[j]+Z[k]]=j*k   # create new partial product
```

Figure 7 *Python algorithm to derive Ludgate's index numbers*
Image reproduced courtesy The John Gabriel Byrne Computer Science Collection

```
final Ludgate Simple Index for each Unit (Table 1):
   0: 50    1:  0    2:  1    3:  7    4:  2    5: 23    6:  8    7: 33    8:  3    9: 14

final Ludgate Complex Index for each Partial Product (Table 2):
   1:  0    2:  1    3:  7    4:  2    5: 23    6:  8    7: 33    8:  3    9: 14   10: 24   12:  9   14: 34
  15: 30   16:  4   18: 15   20: 25   21: 40   24: 10   25: 46   27: 21   28: 35   30: 31   32:  5   35: 56
  36: 16   40: 26   42: 41   45: 37   48: 11   49: 66   54: 22   56: 36   63: 47   64:  6   72: 17   81: 28

final Ludgate Partial Product for each Complex Index (Table 3):
   0:  1    1:  2    2:  4    3:  8    4: 16    5: 32    6: 64    7:  3    8:  6    9: 12
  10: 24   11: 48   12:      13:      14:  9   15: 18   16: 36   17: 72   18:      19:
  20:      21: 27   22: 54   23:  5   24: 10   25: 20   26: 40   27:      28: 81   29:
  30: 15   31: 30   32:      33:  7   34: 14   35: 28   36: 56   37: 45   38:      39:
  40: 21   41: 42   42:      43:      44:      45:      46: 25   47: 63   48:      49:
  50:  0   51:  0   52:  0   53:  0   54:      55:      56: 35   57:  0   58:  0   59:
  60:      61:      62:      63:      64:  0   65:      66: 49   67:      68:      69:
  70:      71:      72:      73:  0   74:      75:      76:      77:      78:      79:
  80:      81:      82:      83:  0   84:      85:      86:      87:      88:      89:
  90:      91:      92:      93:      94:      95:      96:      97:      98:      99:
 100:  0                                                            © Brian Coghlan 2020
```

Figure 8 *Results from running the algorithm of Figure 7*
Image reproduced courtesy The John Gabriel Byrne Computer Science Collection

III.4. Principles of operation of Ludgate's logarithmic slides

Ludgate employed logarithmic index "slides" to implement his Irish Logarithms. The slides had a logarithmic profile representing a simple index on one axis X for each ordinal on another normal axis Y. Figure 9 shows a slide with annotations for the decimal, ordinal and index numbers (based on [38]). It also shows a storage shuttle with a type rod representing the decimal value of '8' (represented by the ordinal value '3'). The principle of operation is that the slide moves right to mate with the shuttle and rod. When it does so, it will have been displaced by a simple index of 3 units. By contrast for a shuttle with a type rod representing the decimal value of '0' (represented by the ordinal value '9'), the slide will be displaced by a simple index of 50 units. It is thought quite likely that ordinals were used in the storage shuttles as a proxy for the decimal operands in order that the series of slide profile changes in Y would be monotonic, as this would facilitate the progressive movement of the slides along the X axis for all possible decimal values.

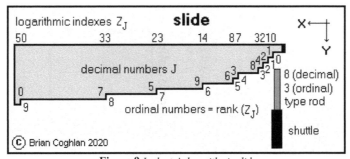

Figure 9 *Ludgate's logarithmic slide*
Image reproduced courtesy The John Gabriel Byrne Computer Science Collection

Note that the slide and shuttle Y-axis of Figure 9 will actually be edge-on, facing directly away from the viewer (reducing their visibility to lines as in Figure 1), but can be illustrated as shown in Figure 10 to aid understanding.

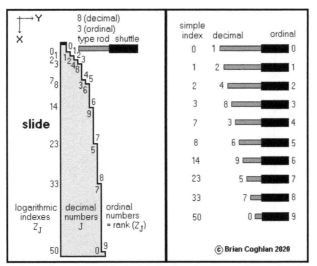

Figure 10 *Re-oriented (left) Ludgate's logarithmic slide, (right) shuttle rod extensions for simple indexes, decimal and ordinal values*
Image reproduced courtesy The John Gabriel Byrne Computer Science Collection

The upper part of Figure 1 and Figure 12 further below show a 6-digit outer shuttle and a 4-digit inner shuttle, representing a machine with a 6-digit outer operand (the multiplicand 813200) and a 4-digit inner operand (the multiplier 9247) as per Engineering [32]. These correspond to the variables to be operated upon, which are from the outer and inner storage cylinders, respectively. Figure 1 and Figure 12 also show six outer slides and one inner slide. The two types of slide are likely to have had the same logarithmic profile, as is shown. Each of the outer slides was to mate with a corresponding digit of the outer shuttle, and were slightly longer (length = maximum compound index) with a pointer attached to one end. The inner slide (length = maximum simple index) was to mate iteratively with one digit of the inner shuttle at a time, starting with the most significant digit.

The shuttles were in line as shown, and the outer and inner slides faced each other on a common 'starting line' aligned with their simple index '0'. To multiply, the slides were moved towards their respective shuttles (i.e. the slides moved towards each other). The fact that the outer and inner slides move towards each other (rather than apart) is a surprise. Figure 11 and Figure 12 show, for example, the leftmost slide when moved to mate with the outer shuttle left type rod representing decimal value '8', and therefore displaced by simple index $Z_8 = 3$. Figure 11 and Figure 12 also show the inner slide when moved to mate with the inner shuttle type rod representing decimal value '9', and therefore displaced by simple index $Z_9 = 14$. Like a slide rule, the relative displacement of the two slides is then the compound index $Z_{72} = Z_{8*9} = Z_8 + Z_9 = 3 + 14 = 17$, representing the decimal partial product $8*9 = 72$.

10

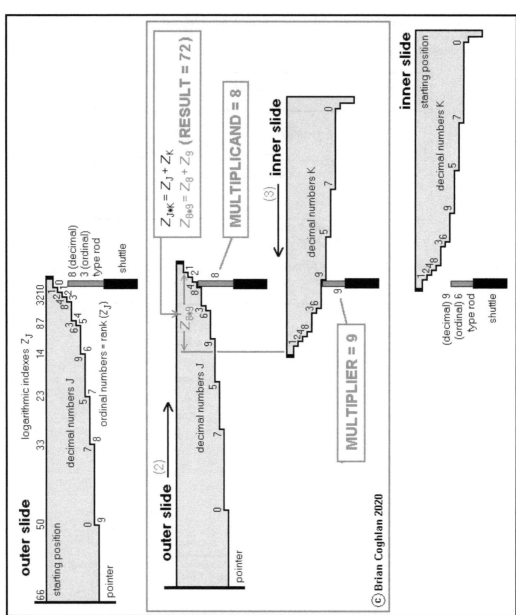

Figure 11 *Multiplication using Ludgate's logarithmic slides:*
Left: outer slide at its starting position,
Right: inner slide at its starting position,
Middle: movement of slides for multiplication of 8*9 = 72
Image reproduced courtesy The John Gabriel Byrne Computer Science Collection

BrianCoghlan-PercyLudgatesAnalyticalMachine-20220320-1305.doc

11

IV. EVIDENCE-BASED DEDUCTION OF HIDDEN ASPECTS OF LUDGATE'S MACHINE

As indicated above, only a few features of his Analytical Machine are described in Ludgate's 1909 paper [1] (***hereafter called Ludgate 1909***), almost everything about its construction is unknown, thus any more must be deduced by contextual analysis of the paper, largely a process of logical inference or elimination of false propositions, and mechanical or mathematical speculations. This document explores that approach, aided by the newly-discovered information of Engineering [32], a copy of which is reproduced as Appendix I.

> **Objective 1: Contextual analysis of the combined texts plus diagram remains questionable, but this document will attempt to show its potential to deduce lemmas and thereby assemble a body of known new design facts.**

The following contextual analysis repeatedly quotes statements from Ludgate 1909 and/or Engineering [32] as evidence to corroborate deductions about the workings of Ludgate's Analytical Machine. This makes the text dense and somewhat turgid, but precise. Where necessary, some speculation will be utilized, but clearly marked as such, in order to explain difficult points.

IV.1. Ludgate's Index mechanism

As mentioned above, Figure 12 reproduces the example multiplication of the outer operand (multiplicand) by the most significant digit of the inner operand (multiplier) of Engineering [32] and also of Figure 1, showing the partial products for all six outer slides. Ludgate's 1909 paper includes an almost identical example. In succeeding iterations the inner slide would mate one multiplier digit at a time with the less significant multiplier digits, while the Mill accumulated the succeeding partial products to produce the result of Ludgate's variant of long multiplication.

Ludgate's index was more than just the slides and shuttles, and in fact was a quite complex multiplicity of moving parts. Much of this related to converting the relative displacement Z_{J*K} to discrete increments or decrements of a result held on the set of counter wheels that Ludgate called the Mill. The two slide types existed within different physical structures. There were twenty of the outer slides (with the pointer attached at one end) in an outer frame but free to move within it (so for illustrative purposes that frame can be treated as invisible). In contrast there was just one inner slide, firmly attached to an inner frame containing multiple blades. There was one blade per compound index, i.e. one per uniquely valid partial product, as per Engineering [32], see Figure 1 or Figure 12. The way in which the blades are composed and utilized is another surprise, but is a very clever arrangement.

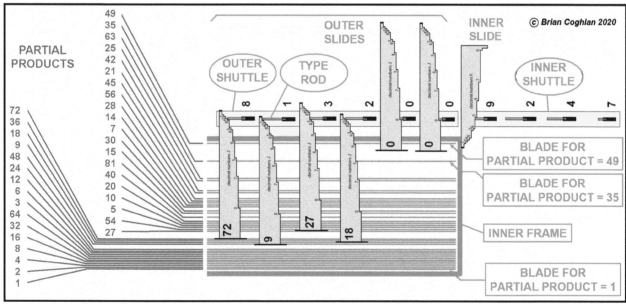

Figure 12 *Ludgate's outer and inner logarithmic index slides, and inner frame with blades for compound indexes, representing the machine with a 6-digit outer operand and a 4-digit inner operand as per Engineering [32] Note that the slides and shuttles will actually face directly away from the viewer, but are shown as is to aid understanding Images reproduced courtesy The John Gabriel Byrne Computer Science Collection*

BrianCoghlan-PercyLudgatesAnalyticalMachine-20220320-1305.doc

It is clear from Ludgate 1909 that the 20-digit multiply-accumulate procedure was as follows (Figure 13 is cut-down example):

(1) The outer (multiplicand b) and inner (multiplier a) operand shuttles are moved to the starting line "near the Index", with the inner slide aligned with the most-significant rod of the inner shuttle (the left rod).

(2) The set of 20 outer slides move in one direction to mate with the shuttle holding the value of the outer operand. The slides convert the digits to "simple index numbers", essentially the logarithmic indexes of the digits of that operand.

(3) Then the single inner slide is moved in the opposite direction to mate with the rod of the shuttle that it is aligned with; the rod holds the value of that digit of the inner operand. The inner slide converts that digit to a simple index number, i.e. the logarithmic index of the value of that digit of that operand.

(4) "as the index is attached to the last-mentioned slide, and partakes of its motion, the relative displacement of the index and each of the [outer operand] slides ... [and] pointers attached to the [outer operand] slides, which normally point to zero on the index, will now point respectively ..." to the compound index number, essentially the sum of the simple index numbers, representing multiplication of the operands.

(5) Next all the compound index numbers are mapped by "movable blades" to numerical units and tens of the partial products and "conveyed by the pointer to" the Mill and accumulated in the Mill.

(6) Then the Index and its attached inner slide performs a "rapid reciprocating action" to align with the rod of the inner shuttle that holds the value of the next less significant digit of the inner operand.

(7) Operations (3-6) are repeated for each of the rods of the inner shuttle "until the whole product of ab is found".

(8) Then "The shuttles are afterwards replaced in the shuttle boxes".

From which may be deduced:

Lemma 4: there is no optimization of shuttle movements, e.g. no early return of the outer shuttle to storage.

Engineering [32] first describes the movement of the outer slides, consistent with step (2) above, but then states "All the slides aforementioned are mounted in a frame, and ... this frame is moved up over another frame divided with another set of index numbers" (**hereinafter called outer and inner frames respectively**). From (3) above it is clear that the inner frame and its slide moves as a whole in the opposite direction to the outer slides.

Lemma 5: the outer slides are mounted on an outer frame, and free to move lengthwise along that frame.
Lemma 6: a single inner slide is mounted to an inner frame, and the whole moves in the opposite direction to the outer slides.

The repetitions of steps of operations (3-6) for each of the rods of the inner operand digits until the whole product is found can be observed in the cut-down 6-digit by 4-digit example in Figure 13. Comparison of the quadrants of Figure 13 very clearly shows how the outer slides do not move as the iterations of operations (3-6) above proceed. Only the inner slide and its frame move together in two dimensions (vertically up to retract, then horizontally right to the next digit, then vertically down to hit that digit's type rod) as they perform each "rapid reciprocating action" of operation (6) above to move digit-by-digit from the most to least significant digit of the inner operand.

13

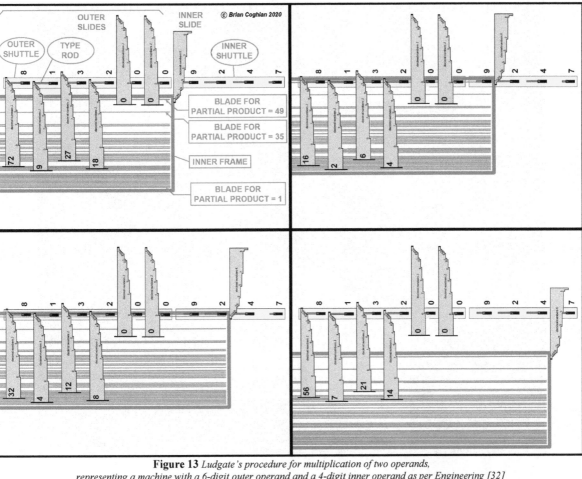

Figure 13 *Ludgate's procedure for multiplication of two operands,*
representing a machine with a 6-digit outer operand and a 4-digit inner operand as per Engineering [32]
Top Left: iteration 1 steps 3-5, multiplicand '813200' multiplied by multiplier digit '9'
Top Right: iteration 2 steps 3-5, multiplicand '813200' multiplied by multiplier digit '2'
Bottom Left: iteration 3 steps 3-5, multiplicand '813200' multiplied by multiplier digit '4'
Bottom Right: iteration 4 steps 3-5, multiplicand '813200' multiplied by multiplier digit '7'
Iteration step 6 occurs between each quadrant of the figure to move the Index and its attached inner slide to the next multiplier digit
Note that the slides and shuttles will actually face directly away from the viewer, but are shown as is to aid understanding
Image reproduced courtesy The John Gabriel Byrne Computer Science Collection

Figure 14 shows an early 3D rendition of the possible re-imagined Ludgate races, shuttles and Index performing the first iteration of the same multiplication as show in Figure 13, i.e. positioned as shown in Figure 12 and in the top-left quadrant of Figure 13. Figure 15 shows a top view (as per Figure 1) of this 3D design performing the iterations over the four least-significant multiplier digits '9247'. Again a principal characteristic is that the inner frame and slide move together in two dimensions with each iteration to reflect the value of the multiplier digits, while the 20 outer slides remain static.

14

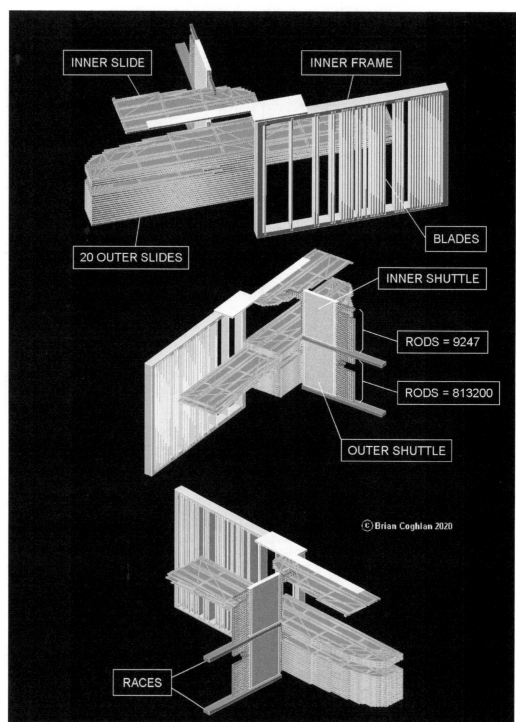

Figure 14 *Early 3D rendition of Ludgate's races, shuttles and Index for multiplication of two 20-digit operands,*
performing the first iteration of the same multiplication as shown in Figure 13,
with multiplicand '813200' multiplied by multiplier '9247' as per Engineering [32]
(the sign rod is at the top of each shuttle, next to the least-significant digit's type rod).
From top to bottom: front, left-rear and right-rear three-quarter views
Image reproduced courtesy The John Gabriel Byrne Computer Science Collection

BrianCoghlan-PercyLudgatesAnalyticalMachine-20220320-1305.doc

15

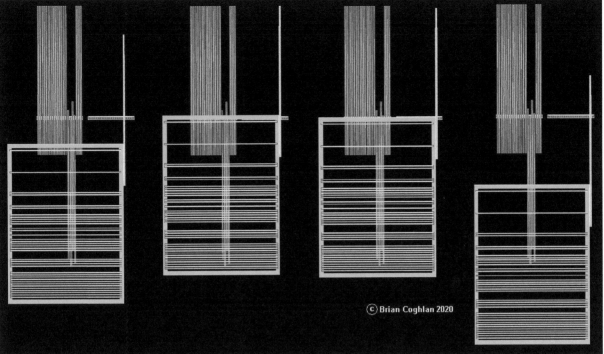

Figure 15 *Early 3D rendition of Ludgate's procedure for multiplication of two 20-digit operands,*
but with multiplicand '813200' multiplied by multiplier '9247' as per Engineering [32]
From left to right: top view of iterations 1 to 4 of steps 3-5 as per Figure 13
Images reproduced courtesy The John Gabriel Byrne Computer Science Collection

Practical design involves issues that are not relevant to operating principles. For example, Figure 16 shows an early 3D rendition of a re-imagining of a slide. It is very thin to limit weight, but has strengthening struts, and has extra height for slots on each side along which it may slide on bars that fit between each slide. The bars are intended to be part of a frame (which is not shown). Each slide is located by two bars on each side (four bars in all), to guarantee physical alignment,

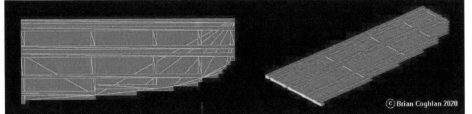

Figure 16 *Early 3D rendition of a re-imagining of Ludgate's logarithmic slide*
Image reproduced courtesy The John Gabriel Byrne Computer Science Collection

Further explanation of more surprising aspects of the Index is greatly helped by the text of Engineering [32] and its diagram, shown in Figure 1 (and in Figure 12 and Figure 13). An important clue is disclosed in Engineering [32], which states "A slide to denote decimal 8 is set at ⅜" above the zero or starting line" (i.e. decimal 8 is represented by its proxy ordinal 3). In this and other statements, Engineering [32] clearly demonstrates a fundamental unit of the machine size is ⅛" (not unexpected for Imperial measurements), that the logarithmic index scale is in units of ⅛" and extends over the length of the largest simple index (50 units, i.e. 6¼"), and that the total allowed movement of an outer slide is 50 units, similarly for the inner slide. When both are at maximum relative displacement, the space occupied will be 100 units, but beyond that the inner frame will extend to the length of the largest compound index ($Z_{7*7} = Z_7 + Z_7 = 33 + 33 = 66$) units, i.e. 8¼", so the maximum space occupied is 166 units, i.e. 20¾". Ludgate 1909 states the machine is "26" long, 24" broad, and 20" high", so the available internal space is just adequate.

Engineering [32] clearly implies the slides embody simple indexes versus ordinals, not decimals, see Table 4.

Decimal operand	Ordinal number	Ordinal profile stop	Simple index	Slide profile stop
0	9	0.000"	50	6.250"
7	8	0.125"	33	4.125"
5	7	0.250"	23	2.875"
9	6	0.375"	14	1.750"
6	5	0.500"	8	1.000"
3	4	0.625"	7	0.875"
8	3	0.750"	3	0.375"
4	2	0.875"	2	0.250"
2	1	1.000"	1	0.125"
1	0	1.125"	0	0.000"

Table 4 *Ludgate's logarithmic slide scales*

Some important spatial relationships on the diagram are not immediately obvious. Mechanically, Figure 1 shows the outer slides (with pointers) are the same length as the inner frame, 66 units, i.e. 16 units beyond the largest simple index. This is because the pointers and blades are used to deduce the partial product. Each blade represents a partial product. The crucial point is that the slides and frame all align along a common starting line, so when idle all the slide tips and the top blade (the partial product line marked "49", where $Z_{7*7} = Z_7 + Z_7 = 33 + 33 = 66$) all lie on the starting line, and the pointers lie above the bottom blade (the line marked "1", where $Z_{1*1} = Z_1 + Z_1 = 0 + 0 = 0$). When an outer slide moves up, its pointer moves up over the blades. When the slide mates with the shuttle, the pointer will lie over a blade that indicates the relevant partial product. The "pointers" are not mentioned in the text or diagram of Engineering [32], but Ludgate 1909 states that "pointers attached to the four slides … now point respectively to the 17th … divisions of the index", i.e. to horizontal lines marked "72 …" that have the round dots at the bottom end of the slides on Figure 1. This confirms that the pointers are at the wide end of the slides, where they lie above $Z_{J*K} = Z_J + Z_K$.

Lemma 7: a fundamental unit of the machine size is ⅛".

Lemma 8: the logarithmic index scales are in units of ⅛".

Ludgate 1909 states that "pointers attached to the four slides, which normally point to zero in the index, will now point respectively to the 17th, 14th, 21st and 15th divisions of the index … corresponding to the partial products 72, 9, 27 and 18" (as in Fig.14 and the top left of Fig.15), and the partial products "are conveyed by the pointers to the mill". So the output is from the wide end of each slide, i.e. from different positions relative to the starting line. There would be great advantage from fixing the position of the Index outputs and Mill so that transfer of value from one to another was convenient, but this is a false proposition. Ludgate does not imply Index outputs are placed in a fixed or convenient place (e.g. the starting line).

Lemma 9: Index outputs are not placed in a fixed or convenient place relative to the Mill.

How do the pointers know the partial product values? On Figure 1, the pointers lie above blades on the inner "frame" attached to the inner "slide", and both diagram and accompanying text of Engineering [32] show there are wide blades at partial product positions, not for 'zero' products, and not everywhere. The blades actually must cover (and be slightly wider than) the span across both sets of shuttle rods. Ludgate 1909 states the blades are movable, and move for a "duration" representing the partial product unit or tens value, "the duration of which displacements are recorded in units measured by the driving shaft on a train of wheels called the mill", so a blade moving "duration" N leads to a pointer moving, which leads to the Mill incrementing by N. The diagram implies (but does not mandate) the pointers lie above the blades, in which case the pointers would be moved or hit by the blade from beneath.

Since pointers are in different positions, several different blade and pointer combinations move simultaneously from different points in space. Figure 1, Figure 12, Figure 13 and Appendices I and III all show just six outer slides with pointers, but in Ludgate's complete machine the Mill has to register the motion of 20 pointers across the horizontal (O for ordinal) axis that are not aligned on the vertical (I for index) axis, and where blade movement can only be towards the viewer on the (P for product) axis, and where blades must move different "durations" for units and tens of partial products. And the combination of positions and movements must be repeated, but with different values for each digit of the inner operand (multiplier). To do this, Ludgate employed coordinate transformation, where a value along one axis is mapped to another value on an orthogonal axis.

Ludgate reduced the multi-dimensional mapping problem by iterating over one inner-shuttle digit number j at a time, and mapping units first then tens. With reference to Figure 1 for each j his approach is equivalent to multiport 2-d mapping combined with 2-d coincident addressing in O and I axes, with the data (partial product units or tens) in the 3rd dimension P. This is very clearly shown when describing the Index as an algorithmic state machine, see Algorithm 1 in pseudo-OCCAM, where SEQ and PAR identify sequential and parallel actions, and ':=' and "=" identify synchronous ("event-triggered" or "clocked") and asynchronous behaviours.

17

PAR *i*=1 to 20	# do simultaneously for all outer operand digits
Outer_S.index(*i*) := map_1(*i*)	# S.indexes(*i*) on *I*-axis ← outer_operand_digit(*i*) on *O*-axis
SEQ *j*=1 to 20	# iterate over inner operand digits
Inner_S.index(*j*) := map_1(*j*)	# S.indexes(*j*) on *I*-axis ← inner_operand_digit (*j*) on *O*-axis
PAR *i*=1 to 20	# do simultaneously for all outer operand simple indexes
SEQ	# but for each outer operand S.index do the following sequentially
C.index(*i*) = Inner_S.index(*j*) + Outer_S.index(*i*)	# C.index(*i*) on *I*-axis = Σ(S.indexes) on *I*-axis
p.product_digit(*i*) := map_2(C.index(*i*, *units*))	# p.product_digit(*i*) on *P*-axis ← C.index(*i*, units) on *I*-axis
p.product_digit(*i*+1) := map_2(C.index(*i*, *tens*))	# p.product_digit(*i*+1) on *P*-axis ← C.index(*i*, tens) on *I*-axis

Algorithm 1 *Pseudo-OCCAM description of Ludgate's Index mappings*
"S.index" refers to simple index and "C.index" refers to compound index

Ludgate 1909 stated a "carriage" moves to select units or tens. Perhaps this "carriage" can jog blades sideways to select units or tens in the form of a stepped blade profile repeating per multiplicand digit. Perhaps slides could "register" partial product units on a "units pointer" on its one side, and then the "carriage" jog the slides sideways to do likewise for tens on its other side, Perhaps there was a separate blade for units and another for tens, although Ludgate 1909 implies this is not so, and the diagram of Engineering [32] only shows one blade per partial product. Just how the "carriage" was utilised remains an open question.

Regarding the blades themselves, some related if disparate facts can be deduced from Ludgate 1909:

(1) "system of slides called the index", i.e. "index" is the whole system of slides, but confusingly in other instances the Index clearly includes the inner frame and blades too, and in other instances seems to be just the inner frame and blades.

(2) "[the index] may be compared to a slide-rule on which the usual markings are replaced by moveable blades", i.e. the blades determine the partial products.

(3) "blades" move for a "duration" representing a partial product digit. This blade motion can only be along the *P*-axis to/from the viewer of the diagram of Engineering [32].

(4) "index is arranged so as to give several readings simultaneously". It can only do that if it is a planar structure as in the diagram of Engineering [32].

(5) "The numerical values of the readings are indicated by periodic displacements of the blades mentioned", possibly "periodic" increments of ⅛".

(6) "In the index the partial products are expressed mechanically by movable blades placed at the intervals shown in column 2 of the third table". Therefore on the inner frame there are moveable blades protruding along the *P*-axis to/from the viewer, one blade for each of 35 non-zero partial products.

(7) "Now the duration of the first movement of any blade is as the unit figure of the partial product which it represents". This strongly indicates that there is only one blade per partial product (not one for units and another for tens), and it first moves along the *P*-axis for units, and later after carriage motion it performs the same function for tens. Ludgate 1909 states "most of the movements … are derived from set of cams placed on a common shaft parallel to the driving shaft". These displacements could be easily implemented by underlying steps or cams, e.g. a double-lobe cam per blade could create the units then tens movements.

(8) "which movements are conveyed by the pointers to the mill". As it is clear the pointers are attached to the slides and move with the slides, this statement is mysterious.

This conveyance from pointer to Mill in (8) above remains the most mysterious aspect of the Index and Mill. Where is the Mill, does it move, and how are Index output values conveyed to it? Engineering [32] says that the partial products are "registered" before "finally being added in the Mill". Hence "pointers" must "convey" or "register" the "duration" or the existence of the "duration", which Ludgate 1909 says must be "recorded in units measured by the driving shaft on a train of wheels called the mill". And yet when blades move, the pointers (which are in different positions) can only move along the *P*-axis to/from the viewer of the diagram of Engineering [32], not towards the Mill. This poses two classes of proposition.

The first "direct" class assumes "duration" means distance, which might suggest each pointer has separate mechanisms that directly act to "register" the partial product units and tens before conveying them, or alternatively that perhaps the "carriage" can directly shift some "conveyance" mechanism, or that the "conveyance" employs mechanisms to directly convey partial products. The blades could be driven by double-lobe cams from below, where units and tens halves have fixed arc-angle raised segments with a variable height proportional to the partial-product digit magnitude, equivalent to the "duration" (like car engine valve cams but with two opposing lobes), see the left side of Figure 17.

The second "indirect" class assumes that "duration" means time, with indirect action, so in the time a blade moves N increments, a driving shaft also indirectly acts to "register" the duration N, and when the blade hits the pointer this action is forcibly enabled or disabled. This would nicely decouple the Mill from the Index. In this case, the blades could be driven by double-lobe cams from below, where units and tens halves have fixed height raised segments over a variable arc-angle

18

proportional to the partial-product digit magnitude, equivalent to the "duration", see the right side of Figure 17.

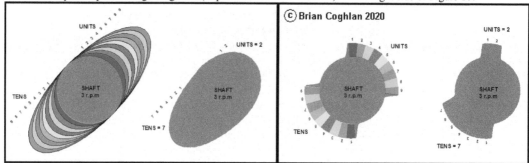

Figure 17 *Left: Double-lobe cam with variable height raised segments (proportional to partial-product digit magnitude) over a fixed arc-angle*
Right: Double-lobe cam with fixed height raised segments over a variable arc-angle (proportional to partial-product digit magnitude)
Image reproduced courtesy The John Gabriel Byrne Computer Science Collection

Either class of propositions would be consistent with both Ludgate 1909 and Engineering [32], but in the absence of further information it has not yet been possible to differentiate further.

Figure 18 shows early 3D renditions of each type of cam (i.e. for each class of propositions) for all the valid partial products. However the blade movements will be preceded by other mechanical movements (e.g. move frame, release slides), and followed by further mechanical movements (e.g. retract slides, move frame to next multiplier digit), so the blade movements will only occupy a portion of the camshaft revolution. Hence in reality the cam lobes will only occupy a portion of the cam revolution, and the cams will need to have a much larger diameter. Other cams on the same shaft, or on a synchronously related shaft, may control the other mechanical movements.

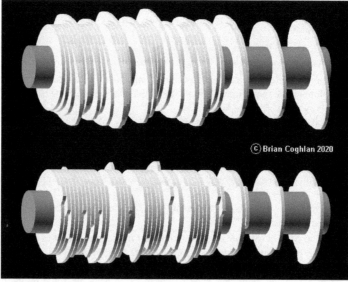

Figure 18 *Top: Double-lobe camshaft with variable height raised segments (proportional to partial-product digit magnitudes) over a fixed arc-angle*
Bottom: Double-lobe camshaft with fixed height raised segments over a variable arc-angle (proportional to partial-product digit magnitudes)
There is one cam per composite index, with the partial product units lobes at the front and tens lobes at the rear
Composite index 0 (partial product 1) is at left, and composite index 66 (partial product 49) is at right
Image reproduced courtesy The John Gabriel Byrne Computer Science Collection

In the absence of detail about the mysterious "conveyance", it is useful to outline three very different speculative but workable solutions that adhere very closely to Ludgate 1909 and Engineering [32]. These cannot be the only possible solutions, but the mechanical details of any other must also adhere.

The first solution assumes "duration" is time, all blades have the same height, blades can independently slide upwards towards pointers, "blade movement" is a fixed distance of variable duration, "register" means to register "enable accumulation", "carriage near Index" is a bar between Index and Mill, and "move from units to tens" means to move the bar from units to tens digits. The blades are moved up by double-lobe cams with fixed height variable arc-angle raised segments, where the angle is proportional to the partial-product digit magnitude (see Figure 17 right). The "carriage" bar moves between units and tens digits in lockstep with the cams, i.e. the bar jogs, not the Index or Mill. When raised by its blade, a pointer activates its mysterious "conveyance", a

19

flexible cable, to stimulate an "enable" lever below the bar to engage escapements or gearing synchronized to the main motor shaft to increment/decrement either its units or tens Mill digit wheel as per the bar position.

An early 3D rendition of this possible "re-imagined" Index is shown in Figure 19, with the Mill assumed to be fixed in position beyond the movable decimal-point carriage bar (at the end of the flexible cables), see further below.

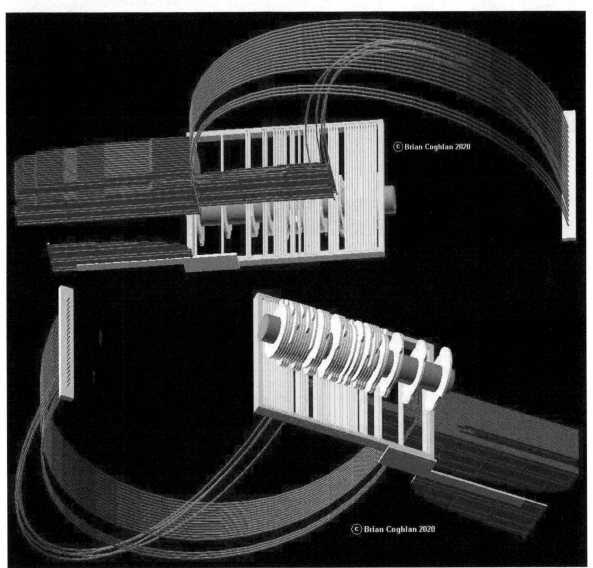

Figure 19 *Early 3D renditions of a possible re-imagined Ludgate Index with flexible cables terminating in decimal-point carriage bar*
Images reproduced courtesy The John Gabriel Byrne Computer Science Collection

As long as the blade is raised, the digit will increment/decrement. "Enable levers" below the decimal-point carriage bar are pushed by the flexible cables to engage a mechanism in one way for increment, and another way with reverse rotation for decrement (as per Ludgate 1909). Figure 20 shows such a reversible escapement mechanism, although reversible gearing is equally applicable. It appears that most mechanical counters use escapement for their first (least significant) stage to guarantee discrete incrementing. In the case of Ludgate's Mill, each wheel is in effect its own first stage, so an escapement may be the most appropriate mechanism.

Brian Coghlan, Published online, 2022. © 2022, The John Gabriel Byrne Computer Science Collection

Percy Ludgate's Analytical Machine

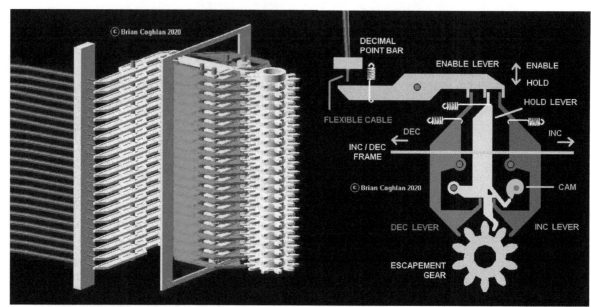

Figure 20 *Speculative "conveyance mechanism using flexible cables to "enable" Mill wheel increment/decrement escapement*
TO BE DONE: *The top layer of escapement levers (included as a placeholder) should be a sign management mechanism*
Image reproduced courtesy The John Gabriel Byrne Computer Science Collection

The mechanism shown in Figure 20 is very speculative. An enable lever keeps increment, decrement and hold levers in the "hold" state until the flexible cable pushes the enable lever into the "enable" state. A snail cam continuously rotates in synchronism with the blade cams to trigger increment or decrement by releasing the hold lever, which itself allows either an increment or decrement lever to strike an escapement gear. Increment or decrement is selected by moving an "inc/dec frame" sideways such that it keeps either the increment or decrement lever in its "hold" state. The enable lever and inc/dec frame are only moved when the cam is in the "hold" region. The increment, decrement and hold levers are spring-loaded as required for escapement. In contrast the enable lever needs only light spring-loading to maintain the hold state. Thus the flexible cables only require light activation force (an important requirement). This escapement mechanism is replicated for each of the 20 Mill digits as shown. A twenty-first escapement gear is included for overflow detection.

This design is only illustrative, as the cam profile, gear lever involution angles, increment and decrement facets, and their relative positioning, have not been formally designed, and the design does not include a carry mechanism nor a sign management mechanism. Nonetheless, eight common problems are overcome. Firstly, if a slide is at index 50 (representing zero) then there is no blade to raise its pointer, so no increment/decrement will occur. Secondly, any partial product that terminates on any part of the decimal-point carriage bar that is beyond the Mill will not be added. Thirdly, any Mill digit that is beyond the decimal-point carriage bar will be unaffected. Fourthly, the motive power for the blades and pointers comes from the blade cams, and that for the Mill comes from the motor, neither comes from the flexible cable "conveyance" (which is driven by the pointers). Fifthly, the flexible cable "conveyance" has a very tolerant binary action: either "enable" or "disable". Sixthly the Mill design is not constrained. Seventhly, the flexible cables allow the Mill to be wider than the Index, which allows a more complex Mill wheel design and also decouples the Mill design from the Index design. Finally, if the flexible cable cores are made of spring steel then the cables provide the motive power for moving the slides once they are released.

The second solution assumes "duration" is length, all blades have the same height, blades can independently slide upwards towards pointers, "blade movement" is a variable distance of fixed duration, "register" means to register "data", "carriage near Index" is a bar between Index and Mill, and "move from units to tens" means to move the bar from units to tens digits. The blades are moved up by double-lobe cams with fixed arc-angle raised segments with a variable height, where the height is proportional to the partial-product digit magnitude (see Figure 17 left). The "carriage" bar moves between units and tens digits in lockstep with the cams, i.e. the bar jogs, not the Index or Mill. When raised by its blade, a pointer activates its mysterious "conveyance", a tooth to jam a ratcheted rack that is driven by gearing from the main motor shaft to increment or decrement either the units or tens Mill digit wheels, see Figure 21.

TO BE DONE

Figure 21 *Speculative "conveyance mechanism using a tooth to jam a ratcheted rack to halt Mill wheel increment/decrement*
Image reproduced courtesy The John Gabriel Byrne Computer Science Collection

21

Until the blade raises the pointer high enough to jam the rack, the digit will increment/decrement. The rack engages gearing for increment, and other gearing with reverse rotation for decrement (as per Ludgate 1909). Similar common problems are overcome by this approach, e.g. if a slide is at index 50 (representing zero) then there is no blade to raise its pointer (so no increment/decrement will occur), nor is the Mill design constrained.

The third but very different solution assumes "duration" is length, all blades are fixed in the inner frame, the whole inner frame with inner slide and blades moves upwards towards pointers, "blade movement" is a variable distance of fixed duration, "register" means to register "data", "carriage near Index" is what positions the Index relative to the Mill, and "move from units to tens" means to jog the whole inner frame with inner slide and blades sideways from units to tens digits. The blades have a stepped profile representing partial product units and tens in a repeating pattern across the blades, i.e. the units and tens steps each have their own variable height proportional to the partial-product digit magnitude, equivalent to the "duration". The whole frame is moved up by a double-lobe cam with two identical fixed arc-angle raised segments of height proportional to the maximum partial-product digit magnitude (see Figure 17 left). The "carriage" jogs the inner frame between units and tens digits in lockstep with the cam, i.e. the Index jogs relative to the Mill. When raised by its blade, a pointer activates its mysterious "conveyance" (e.g. either of the conveyances described above) to increment/decrement the Mill. One issue is that the jogging and upward movements of the inner frame decouple the inner slide from the inner shuttle, so the frame will need to be locked in its position on the I-axis while the movements take place. An interesting variant would employ separate units and tens pointers that directly 'register' the blade displacements on ratcheted racks along the pointers (with blade profiles ensuring only one pointer moves at a time), so that when the twenty outer sliders are returned to their starting position they physically 'convey' the pointers to engage with the Mill, where the displacements are reversed, thereby incrementing/decrementing the Mill – all this would be challenging on a ⅛" pitch, but might resolve some mysteries regarding the Mill (see Section IV.3) and carry propagation (see Section IV.6).

All three are entirely speculative solutions to an absence of any relevant detail in Ludgate 1909 or Engineering [32].

IV.2. Ludgate's special mechanism for ordinals

Ludgate 1909 says: "ordinals are not mathematically important, but refer to a special mechanism which cannot be described in this paper" (why? patent application?). Ordinals simply rank the logarithmic indexes in increasing order, but the statement implies they are used in some way. Ludgate gives no indication that the shuttles hold ordinals rather than ordinary integers, but Riches (University College Swansea) [42] assumes the shuttles hold ordinals, whilst McQuillan [38] both assumes this and discusses why. Ludgate does make clear that the Mill wheels hold decimal values, so there are two obvious possibilities:

(1) Shuttles hold decimal numerical values, so write-back would be unmapped from Mill to shuttles for Storage, and the special mechanism would convert the shuttle numerical values to ordinals for input to the Index. This would need two converters, one each for outer and inner shuttle, but would allow a fanout from the shuttles to a wider Mill.

(2) Shuttles hold ordinals, so write-back would be mapped via the special mechanism to convert Mill numerical values to ordinals for Storage. This would need just one converter at the output of the Mill.

Simplicity would favour option (2), i.e. shuttles hold ordinals.

How then could the special mechanism convert from the numerical values to ordinals? Figure 4 on page 76 of Riches report [42], an excerpt of which is shown in Figure 22, is instructive. It proposed that to store a result from the Mill, a shuttle (presumably with all rods extended) should be pushed to mate with notched sliders displaced according to the Mill wheel rotations (perhaps via racks). Shuttle rods would then be backed into the shuttle according to notch depths, so that Mill wheel rotations would map to notched slider displacements, which would then map to notch depths proportional to result values, where the latter mapping could be to ordinals. In this case some further mechanism would be needed to allow Ludgate's Divide and Log cylinders to store ordinal results directly into shuttles, or perhaps to engage with the notched sliders as needed to achieve this.

This is a viable solution, but could easily be extended to fully integrate the Divide, Log, and any other conceivable cylinders. Ludgate 1909 states a cylinder value "comes opposite to a set of rods. These rods then transfer that number to the proper shuttle" – these "rods" could be Riches' notched sliders. This could allow integration: each Mill wheel could drive other cylinders, e.g. an extra cylinder[1] with values representing the Mill contents, alongside the Divide and Log seeding cylinders, all sharing the "special mechanism" between the Mill and Storage (see Section IV.3 for how this might resolve other mysteries). As per Ludgate 1909, the 20-digit Divide and Log cylinders would be addressed with the three most-significant Mill digits (see Section IV.8) – the example extra cylinder would need to comprise either 20 cylinders, each independently addressed by its Mill digit, or perhaps 7 cylinders of 0-999, each addressed by a different subset (triple) of the twenty Mill digits (the latter solution would ensure all cylinders had identical diameters and mechanisms, although 0-99 of the Divide and Log cylinders would be unused). All cylinders could output results to a shuttle via the "special mechanism", which would realize mapping to ordinals. Alternatively that mapping could be done by the cylinders, with hole depths proportional to ordinals and the notched sliders realizing identity mapping. In either case   Figure 22 shows the resulting path via which such cylinders would in effect write ordinals to shuttles.

---

[1] In previous versions of this document this was called an "Ordinal" cylinder, a potential misnomer.

Therefore let us propose that the formats of the numbers in the shuttles, Index and Mill are ordinals, simple/compound indexes, and ordinary integer numerical values respectively, and that all cylinders engage the special mechanism to convert numerical values to ordinals. The propositions need to be carefully scrutinized, as clearly Ludgate could have intended other arrangements.

Proposition 1: Let us propose numbers in Storage are ordinals.
Proposition 2: Let us propose Ludgate's "special mechanism" is implemented for any cylinder to convert from decimals to ordinals.
Proposition 3: Similarly, let us propose Ludgate's Divide, Log and any other cylinders are addressed in decimal.

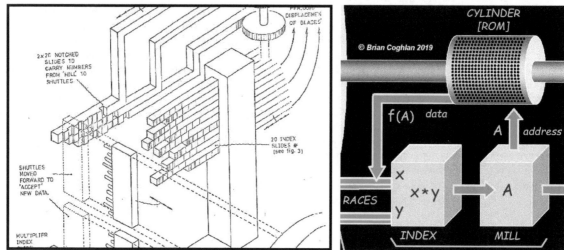

Figure 22 *Left: Figure 4, page 76, of Riches report, showing shuttle rods pushed to mate with notched sliders*
Image reproduced courtesy Professor John Tucker, University College Swansea, UK
Right: Conjectural diagram of path for cylinders to write ordinals to shuttles
Image reproduced courtesy The John Gabriel Byrne Computer Science Collection

IV.3. Ludgate's Mill

More complex Mills could not fit within ⅛", but might within ¼" or ⅜". This highlights that it might be convenient to base the Mill on a wider pitch than ⅛" (especially to propagate carries, address Divide, Log and any other conceivable cylinders, and support any related actions).

The mystery of conveyance from pointers also impacts the Mill. Ludgate 1909 gives little detail:

(1) "a carriage near the index now moves one step to effect multiplication by 10". He doesn't say the "mill carriage", so it seems the Mill does not move, instead a separate carriage moves.

(2) "the duration of [blade] displacements are recorded in units measured by the driving shaft on a train of wheels called the mill". As discussed before, this might suggest the Mill is incremented by the driving shaft, not by the pointers, i.e. "duration" is time.

But Engineering [32] includes new and puzzling details of how the Mill is affected:

(3) For the 1st digit "9" of the multiplier "All these partial products are registered in the mill below, as indicated", and "In a final operation these partial products are added together as indicated, giving 7,318,800". Does this "final operation" imply racks or wheels that "register" partial products, and that the "final operation" adds these to further "partial-sum wheels"? Or does it just mean "finally the result is 7,318,800"?

(4) For the 2nd digit "2" of the multiplier "These partial products will then appear on the mill and be added together, giving the result of the multiplication of 813,200 by 2" (not 7,318,800 + 813,200*2). Does the absence of accumulation mean a set of "partial-sum" wheels for each multiplier digit?

(5) Finally "The process is repeated for the remaining figures of the multiplier, and the whole added together so as to give the product of 813,200 x 9247". Does this mean yet another "final-product" set of Mill wheels? Or again does it just mean "finally the result is 813,200 x 9247"?

These sentences appear to be in contradiction with Ludgate 1909, which seems to imply that the partial products are accumulated on just one set of Mill wheels. What does "register" mean? Mechanically, it may mean to physically "convey" the partial products to the Mill, with jogging right then left to add units then tens, perhaps via gears on either side of Mill wheels, or via "open" or "spur" differentials [43]. Alternatively, "register" may reflect a different use of language, e.g. to register an "enable" by activating a mechanism to allow accumulation by the Mill as in Figure 20 or Figure 21.

23

As a further alternative, "register" could mean transfer to one set of Mill wheels, but accumulation on a second "final-product" set of Mill wheels as per (3) and (5) above. Ludgate's use of cylinders as lookup tables would allow them to double-act as this second stage of the Mill, rotating to accumulate in decimal (eliminating "addressing" of cylinders), and resolving contradictions. It would be physically efficient if cylinders were grouped as 7 cylinders of 0-999, each accumulating for a different triple of the twenty Mill digits, consistent with suggestions in Section IV.2 above; this would provide a ⅜" rather than ⅛" pitch for the mechanism, and reduce carry mechanisms by a factor of three. The penalty would be increased accumulation time (counting 0-999 rather than 0-9), which would contradict Ludgate's timing claims, so mystery remains.

In addition, the term "register" may be related to handling decimal-points. If numbers have a decimal-point, then each multiply result decimal-point will depend on its operand decimal-points, and may differ from the accumulated Mill decimal-point. So the result may need to be aligned to the Mill decimal-point before accumulation.

IV.4. Ludgate's data format

Ludgate's numbers used a sign-magnitude format. Figure 1 shows an 'x' beyond the least-significant end of the digits of the outer frame. This can only be the sign, and therefore the sign is stored beyond the least-significant rather than the most-significant digit. Given multiplication begins at the MS digit, this may be mechanically significant.

Lemma 10: sign is stored beyond the least-significant digit.

Were numbers left or right justified? For the case where S.index(8132)=3071 and S.index(0)=50, Ludgate 1909 stated that "the first four slides will therefore move 3, 0, 7, and 1 units respectively, the remainder of the slides indicating zero by moving 50 units." Does "first four" imply numbers are right-justified? The answer is no, since if the "remainder" was all leading zeroes, why say "first four", as a fifth would not exist.

Ludgate 1909 also stated "Another slide moves … to the simple index number of the first digit of the multiplier". This (and more) indicates long multiplies started with the multiplier left digit. But does that imply numbers are left-justified? The answer in this case is maybe. But there is a problem with left-justification, as then the decimal point position must be known, otherwise "1" could mean any of 10^0 to 10^{20}.

There are only three remaining possibilities: (a) numbers are floating-point, (b) numbers are fixed-point, and (c) numbers are left-justified. Which of these did Ludgate employ? A floating-point format is likely to have been far too ambitious for the time (although Torres y Quevedo published a paper design in 1911 [7]). However, Ludgate said: "the position of the decimal point in a product is determined by a special mechanism which is independent of both mill and index." This suggests any implied decimal point position may be capable of being varied, and that it was an issue for Ludgate. But he also said "rods for every figure of the Variable, and one to indicate the sign", i.e. only digits and sign were stored in a shuttle. No rod was used for decimal point. Without rods for decimal point, floating-point or left-justified numbers cannot be used. Hence numbers must be fixed-point, with an implied decimal-point placement.

Lemma 11: numbers must be fixed-point

This suggests the decimal point in operands and results would be known to the programmer. The diagram of Engineering [32] shows a multiplicand of "813200", without indicating an implied decimal-point. It could be after digit 6, or as an arbitrary example, say after digit 18:

$$+813200.00000000000000 \text{ or } +000000000000813200.00$$

In the limit, allowing an implied decimal-point provides an exponent range as follows:

$$\pm 0.0000000000000000001 = \pm 1.0 \times 10^{-19} \text{ to } \pm 10000000000000000000 = \pm 1.0 \times 10^{+19}$$

Potentially the decimal-point could be controlled or managed from the upper keyboard (like the IBM 360/44 floating-point precision could be varied with a control-panel selector [44]), but Ludgate 1909 stated the number-paper and upper keyboard are synonymous, so this implies any control was via an instruction that could also be read from the number paper.

IV.5. Ludgate's decimal-point alignment

Wherever the decimal-point is, since divide and log operations rely on multiply, add and subtract operations, and the latter two preserve any uniformly implied decimal-point, then multiply may be the only operation for which the decimal-point must be aligned. In most accumulator architectures this is done via shift instructions, but this appears to be excluded by Ludgate's statement that decimal-point position is "determined independently of these [divide and log] formulae", i.e. those formulae (which use multiply intensively) neither know nor care about the decimal-point.

He also states "the position of the decimal-point in a product is determined by a special mechanism which is independent of both mill and index". This suggests a separate mechanism that can be set to determine the decimal-point for subsequent operations, and that its impact is on the product, not the Mill. The special mechanism would need to be set statically or dynamically as required by the programmer. It would then align subsequent products with the accumulated result.

Percy Ludgate's Analytical Machine

24

By way of example, for multiplying 6-digit numbers with the decimal-point in the middle:

$$005.210 * 002.710 = 000014.119100$$

This can be represented as integer multiplication followed by scaling (shifting right by 3 digits):

$$005210 \times 002710 = 000014119100 / 1000000 = 000014.119100 \rightarrow 014.119$$

Within the special mechanism, scaling is most easily done by shifting. There are two reasonable options:

a) Pre-shift with a 20-digit product or wider beyond the least-significant digit for more accuracy: repeated shifting during multiply is required, but overflow detection is easy.

b) Post shift with a 40-digit product: overflow detection is harder, but shifting can be a subsequent action.

Any alignment operation is likely to have been designed to allow left or right shifts by $1 < N < 20$ digits.

Lemma 12: the "special mechanism" is likely to support an alignment operation.

A 40-digit product as per (b) raises other issues, e.g. rounding (but Ludgate 1909 does not mention it), or the fact that Ludgate only allowed 20 time-units for the final ripple carry (but it _is_ feasible to do _only_ 20-time-units of the ripple carry).

The early 3D rendition of a "re-imagined" Index shown in Figure 19 includes a bar as the special mechanism. The bar acts on an "enable" lever below it to engage a gear (per digit) from the main motor shaft, so as long as a blade/pointer is raised, its digit(s) will be enabled to increment/decrement. To change the decimal-point position, here the slides, blades, pointers and Mill do not change position, only the bar shifts left/right, but since the principle is relative alignment between bar and Mill, the bar might equally well be fixed and the Mill shifted. An instruction shifts the bar versus Mill so that only the digits that should for that decimal-point setting are affected. This mechanism supports non-accumulation of zeroes, and also "unaffected digits", where depending on the alignment, some Mill digits will be beyond the bar, and so the enable lever for those "unaffected" digits will not move. This workable solution adheres very closely to Ludgate 1909. Again this cannot be the only possible solution, but any others must also adhere.

Alignment might also explain Ludgate's choice to multiply in reverse to the traditional order (with its awkward step left then two steps right mentioned in Section III.1), as it may have been an optimization to allow early termination of multiplication once the partial products became less significant than the Mill's least-significant digit (but his fixed timing for multiply implies this was not done), and/or to allow skipping zeroes more quickly, and/or to facilitate carry propagation.

IV.6. Ludgate's carry propagation

Ludgate said he had a new way to do carrying which decoupled adding and carrying, and he stated that the "carrying is practically in complete mechanical independence of the adding process, so that the two movements proceed simultaneously", that "the sum of m numbers of n figures would take $(9m + n)$ units of time" ("n" represents a final ripple carry at the end of a sequence of m additions), and for multiply, the $n = 20$ digit multiply-accumulate involving $m = 40$ additions would take $((9 \times 40) + 20) = 380$ time-units. If carrying is serialized, multiply would take 420 time-units, therefore his timings suggest synchronous pipelining of ripple carries, overlapped with mechanical movements, 30 years before the simple use of pipelining in Zuse's Z1 and Z3 computers and 60 years before its use in supercomputers [45].

Ludgate's multiplier inner slide and frame proceeds from MS to LS multiplier digit. For any multiplier digit N, after the partial products units register/add there is mechanical movement of a "carriage near the Index" in order to switch to tens. If units carry was synchronously latched, then its propagation could be overlapped with this movement. After the partial products tens registers/adds, the movements back to the units digits occurs, and if their tens carry were synchronously latched then their propagation could be likewise overlapped. In each case (sum + carry) is always less than 19 so double-carries (i.e. >19) are avoided, in contrast to if units and tens carries were propagated together. The "carriage" therefore gives each digit two opportunities to propagate a carry, allowing up to two synchronous carries overlapped ("hidden") behind carriage movements, and optionally another (highly unlikely) one behind the inner slide and frame movement to the next LS digit.

Proposition 4: synchronous ripple carry propagation is overlapped with mechanical carriage movements, effectively taking zero time-units.

The corollary is that the time allowed for these carriage movements must be long enough for the carry to increment the wheel.

Proposition 5: time allowed for the mechanical carriage movement is one time-unit.

A counter-argument is that if the two movements do _actually_ proceed simultaneously, then Ludgate's solution was different, for example employing a differential in the Mill or decimal-point "special mechanism" as mentioned above, or even a Mill with a differential and second set of wheels.

IV.7. Ludgate's Mill microarchitecture

The data format. alignment, and carry propagation all impact on the Mill mechanism. Its microarchitecture is very clearly shown when describing a <u>single digit</u> of the Mill as an algorithmic state machine, for example as per the speculative pseudo-OCCAM Algorithm 2 below, which assumes Proposition 4 and Proposition 5 directly above are true, and handles both positive and negative numbers, but ignores decimal-point alignment. Table 5 shows its expected behaviour. This demonstrates how within each multiplier iteration, each digit can allow two separate synchronous carries overlapped ("hidden") behind the separate units and tens carriage movements, hence ensuring there is never a carry of two, as outlined in the previous section.

Mechanically, "Wheel" can be each "figure-wheel" (digit wheel) of Ludgate's Mill, where "Wheel = [9 | 0]" signifies the typical lug between "9" and "0" on the wheel, "CYout" is a typical mechanical latch, and ":=" the starting or active transition of Ludgate's "time-unit being the period required to move the figure-wheel through $1/10$ revolution". For negative numbers CYin and CYout represent borrow input and output.

```
SEQ  i = 1 to n                                          # n is the number of digits (20)
    ################### UNITS #################
    PAR
        moveTo ( Units );                                # mechanical "carriage" movement to units
        IF ( CYin = 1 )                                  # overlap carry/borrow propagation if necessary
            IF ( POSITIVE ) PAR
                IF ( Wheel = 9 ) CYout := 1 ELSE CYout := 0;   # latch carry output as needed
                Wheel := Mod10 ( Wheel + 1 ) ;           # hidden pipelined ripple carry propagation
            ELSE PAR
                IF ( Wheel = 0 ) CYout := 1 ELSE CYout := 0;   # latch borrow output as needed
                Wheel := Mod10 ( Wheel - 1 );            # hidden pipelined ripple borrow propagation
        SEQ p = 1 to partialProduct ( Units )            # iteratively add partial product units (inc/dec wheel)
            IF (POSITIVE ) PAR
                IF ( Wheel = 9 ) CYout := 1;             # latch carry output as needed
                Wheel := Mod10 ( Wheel + 1 );            # add partial product units
            ELSE PAR
                IF ( Wheel = 0 ) CYout := 1;             # latch borrow output as needed
                Wheel := Mod10 ( Wheel - 1 );            # subtract partial product units
    ################### TENS #################
    PAR
        moveTo ( Tens );                                 # mechanical "carriage" movement to tens
        IF ( CYin = 1 )                                  # overlap carry/borrow propagation if necessary
            IF ( POSITIVE ) PAR
                IF ( Wheel = 9 ) CYout := 1 ELSE CYout := 0;   # latch carry output as needed
                Wheel := Mod10 ( Wheel + 1 );            # hidden pipelined ripple carry propagation
            ELSE PAR
                IF ( Wheel = 0 ) CYout := 1 ELSE CYout := 0;   # latch borrow output as needed
                Wheel := Mod10 ( Wheel - 1 );            # hidden pipelined ripple borrow propagation
        SEQ p = 1 to partialProduct ( Tens )             # iteratively add partial product tens (inc/dec wheel)
            IF (POSITIVE ) PAR
                IF ( Wheel = 9 ) CYout := 1;             # latch carry output as needed
                Wheel := Mod10 ( Wheel + 1 );            # add partial product tens
            ELSE PAR
                IF ( Wheel = 0 ) CYout := 1;             # latch borrow output as needed
                Wheel := Mod10 ( Wheel - 1 );            # subtract partial product tens
    ################### FINAL RIPPLE CARRY #################
    SEQ k = 1 to n                                       # n is the number of digits (20)
        IF ( CYin = 1 )
            IF ( POSITIVE ) PAR
                IF ( Wheel = 9 ) CYout := 1 ELSE CYout := 0;   # latch carry output as needed
                Wheel := Mod10 ( Wheel + 1 );            # final pipelined ripple carry propagation
            ELSE PAR
                IF ( Wheel = 0 ) CYout := 1 ELSE CYout := 0;   # latch borrow output as needed
                Wheel := Mod10 ( Wheel - 1 );            # final pipelined ripple borrow propagation
```

Algorithm 2 *Example speculative pseudo-OCCAM description of a <u>single digit</u> of Ludgate's Mill*

Example	Before mechanical carriage movement			After mechanical carriage movement			After iterative addition	
	CYin	Wheel	p.Product	Wheel	p.Product	CYout	CYout	Wheel
1	0	4	4	4	4	0	0	8
2	1	4	4	5	4	0	0	9
3	0	4	5	4	5	0	0	9
4	1	4	5	5	5	0	1	0
5	0	9	0	9	0	0	0	9
6	1	9	0	0	0	1	1	0
7	0	9	9	9	9	0	1	8
8	1	9	9	0	9	1	1	9

Table 5 *Example expected behaviour of a <u>single digit</u> of Ludgate's Mill as described by speculative Algorithm 2*

IV.8. Ludgate's division and logarithm algorithms

The way in which Ludgate performed division was very novel for 1909. Seeded with an approximate value A of the reciprocal of the divisor q, the algorithm successively converged to the correct answer, using just multiply/accumulate. Using the Binomial Theorem he expanded division to:

$$\frac{p}{q} = \frac{A\,p}{A\,q} = \frac{A\,p}{1+x} = A\,p\,(1+x)^{-1} = A\,p\,(1 - x + x^2 - x^3 + x^4 - x^5 + x^6 - x^7 + x^8 - x^9 + x^{10}) + etc \quad .... \text{[A]}$$

$$= A\,p\,(1-x)(1+x^2)(1+x^4)(1+x^8) + etc \quad .... \text{[B]}$$

The seed A is a 20-digit value taken from a lookup table that contains all the reciprocal values $A = \frac{1}{f}$, where ($100 < f < 999$) is equal to $q[1..3]$, i.e. digits 1-3 of q. As a result A will be slightly larger than q, and therefore $A\,q = 1 + x$, where x is a small fractional value. This relation can be reversed to yield $x = A\,q - 1$ in order to evaluate the series above. Ludgate 1909 states that series A up to x^{10} and series B up to x^8 converge to the value of $(1 + x)^{-1}$ correct to at least twenty figures.

The statement that "the quantity A must be the reciprocal of one of the numbers 100 to 999", i.e. f = 100-999, is a clue. For fixed-point it would be expected that f = 000-999, whereas 100-999 clearly indicates the divisor was left-justified, not fixed-point, having excluded divide-by-zero beforehand.

Ludgate 1909 states "the 900 values of A are stored on a cylinder—the individual figures being indicated by holes from one to nine units deep on its periphery", addressed by $q[1..3]$. The row of holes for A "comes opposite to a set of rods. These rods then transfer that number to the proper shuttle, whence it becomes an ordinary Variable". This means the store instruction either must have a "source operand" that specifies that its data comes from the Divide, Log, or any other conceivable cylinder, or there must be an instruction variant for each source. See Section IV.11 below for a discussion about this.

Ludgate 1909 then states A is "used in accordance with the formula" where a "dividing cylinder, on which this formula is represented in the proper notation of perforation" within the machine's sequencer contains the algorithm to evaluate the series. This means A must be stored in a shuttle reserved for use by the division formula, and that A must either be preserved across or precluded from preemption. It also means the division formula uses the standard instruction perforations, so it is a built-in subroutine like DEC AlphaPALcode [46] rather than microcode (which would use a lower-level of micro-instructions).

According to Boys 1909 the machine first calculates the value of the series, then multiplies the result by $A\,p$. Including the lookup of A, the division might be performed as per Algorithm 3, with steps (6-8) calculated at a lower level as per Algorithm 4.

---

(1) Load q into the Index or Mill
(2) IF (q = 0) THEN error
(3) Left-justify q
(4) Extract the 3-digit value ($100 < f < 999$) = $q[1..3]$
(5) Transfer the 20-digit value A = lookup($\frac{1}{f}$) from the DIV cylinder to storage
(6) Calculate $x = (A\,q - 1)$ and its powers
(7) Calculate the series $S = (1 - x + x^2 - x^3 + x^4 - x^5 + x^6 - x^7 + x^8 - x^9 + x^{10})$ .... [C]
   or possibly the series $S = (1-x)(1+x^2)(1+x^4)(1+x^8)(1+x^{16})$ .... [D]
(8) Calculate the final result $\frac{p}{q} = A\,p\,S$

Algorithm 3 *Ludgate's division algorithm*

27

10 MUL (*A, q*)	# calculate & store *A q*		54 MUL (*-x, +1*)	# load *-x*
11 STO (*Aq, -*)			57 MAC (*x2, +1*)	# calculate $-x + x^2$
14 MUL (*+1, +1*)	# calculate & store *-x*		60 MAC (*-x3, +1*)	# calculate $-x + x^2 - x^3$
17 MAC (*Aq, -1*)			70 MAC (*-x3, -x*)	# calculate $-x + x^2 - x^3 + x^4$
18 STO (*-x, -*)			73 MAC (*-x5, +1*)	# calculate $-x + x^2 - x^3 + x^4 - x^5$ & store
28 MUL (*-x, -x*)	# calculate & store x^2		74 STO (*tmp, -*)	
29 STO (*x2, -*)			84 MAC (*tmp, -x5*)	# calculate $-x + x^2 - x^3 + x^4 - x^5 + x^6 - x^7 + etc$
39 MUL (*x2, -x*)	# calculate & store $-x^3$		87 MAC (*+1, +1*)	# calculate $S = 1 - x + x^2 - x^3 + x^4 - x^5 + x^6 - x^7 + etc$
40 STO (*-x3, -*)			88 STO (*tmp, -*)	
50 MUL (*-x3, x2*)	# calculate & store $-x^5$		98 MUL (*p, tmp*)	# calculate *p S*
51 STO (*-x5, -*)			99 STO (*tmp, -*)	
			109 MUL (*A, tmp*)	# calculate final result *A p S*

Algorithm 4 *Possible low-level in-line implementation of steps (6-8) of division Algorithm 3*
The numbers before the instructions show the estimated timeline in seconds
Instruction types as speculated in Section IV.11

This approach is common in modern computing, although the algorithms are different. Ludgate's seeds were stored on a special cylinder addressed by a subset of the contents of the Index or Mill, essentially a read-only memory used as a lookup table. He designed for a seed lookup table with 900 entries, enough to possibly require a cylinder of the same size as the storage cylinders (although a spiral mechanism might alleviate this). McQuillan [38] discusses the very significant difficulties of Ludgate's approach in some detail. A simpler prototype design could use 90 (or even 9) seeds and still work as he proposed, but converge more slowly. Ludgate 1909 states that division takes 90 seconds, whereas if addition is assumed to take 3 seconds, multiplication takes 10 seconds, and store takes 1 second, then the un-optimized Algorithm 4 takes 109 seconds.

Why did Ludgate not use logarithms for division? The reason is his logarithmic multiplication was mechanically unidirectional. It was done multiplier digit by digit, to yield a 2-digit partial result, before it moved on to the next digits. To do division another mechanism would be needed to take 2 digits of the dividend and 1 digit of the divisor, to yield a 1-digit partial result. While it could be done, it was mechanically much more efficient to have only one mechanism. Ludgate's divide by convergent series still had much better performance than division by iterative subtraction.

Ludgate also "extended this system to the logarithmic series … a logarithmic cylinder which has the power of working out the logarithmic formula, just as the dividing cylinder directs the dividing process. This system of auxiliary cylinders and tables for special formulae may be indefinitely extended." He does not specify the calculation of logarithms, but for example, the natural logarithm might have been calculated using the MacLaurin expansion:

$$ln(1+x) = x - \frac{x^2}{2} + \frac{x^3}{3} - \frac{x^4}{4} + \frac{x^5}{5} + etc$$

Ludgate 1909 states the logarithm calculation takes 120 seconds.

IV.9. Ludgate's storage

Ludgate's arrangement for storage was novel and a major advance on anything before. Numbers were represented as a rod for sign and for each of 20 digits, with every digit protruding 1-10 units. For example, these could be square rods with notches, with some retention. These 21 rods were stored in shuttles, see Figure 23. Two concentric cylinders (one inner and one outer) held the shuttles, i.e. the inner cylinder was inside the outer cylinder. There were a total of 192 shuttles. Essentially he proposed either two linearly addressed memory partitions, or in a single linearly addressed memory partition.

How were the shuttles arranged? Let us assume the shuttles are all the same size, and occupy the same slice width W at their cylinder inner-radius. If inner and outer cylinder inside-radii are A and B respectively, cylinders hold $N = \frac{2\pi A}{W}$ and $M = \frac{2\pi B}{W}$ shuttles each, and $N + M = 192$, then $N = \frac{192 A}{A + B}$. Each shuttle holds 21 rods, and if each is the fundamental unit of the machine size of $^1/_8$", then the minimum height of shuttle is $(B - A) = 2^5/_8$", realistically *3*", i.e. the outside-diameter of the outer cylinder is then *(2A + 12)*". Ludgate 1909 states the machine is machine is 26" long, 24" broad, and 20" high. Therefore *(2A + 12) < 20*, yielding a maximum *A = 4"* and *B = 7"*, which sadly leads to an irrational $N = \frac{192 A}{A + B} = 69.8181...$ But if *A = 3"* and *B = 6"*, then *N = 64, M = 128, M = 2N, B = 2A*, and the outside-diameter of the outer cylinder is 18". This simple and very convenient geometric result very strongly suggests Ludgate designed for 64 inner and 128 outer shuttles, most easily addressable in binary.

Lemma 13: storage is arranged as 64 inner and 128 outer shuttles.

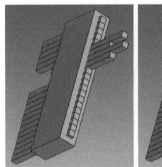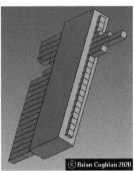

Figure 23 *Speculative schematic of shuttles storing +813200.00 as ordinal +304199.99 (left) and +9167.00 as ordinal +6058.99 (right).*
The sign is the rod at the top, the next rod is the LS digit, and the bottom rod is the MS digit.
The stored profiles are in reverse ranking of ordinals, see Table 4.
The shuttle housing and rods are to correct scale but not representative in any other way..
Images reproduced courtesy The John Gabriel Byrne Computer Science Collection

Ludgate 1909 stated that to access shuttle S, the appropriate cylinder must be rotated to line up S with the Index input race. He also explained that if an inner shuttle held the value X and an outer shuttle held the value Y, then to multiply $X * Y$ the inner cylinder must be rotated to shuttle X and the outer cylinder rotated to shuttle Y, then the Index instructed to multiply $X * Y$, then the Mill instructed to accumulate the result, and finally restore the shuttles to their cylinders. The result could then be stored back simultaneously to up to two shuttles. Ludgate did not state where these two shuttles were allowed to be, only that "they do not belong to the same shuttle-box", where there are "two coaxial cylindrical shuttle boxes". Clearly one must be an inner and one an outer shuttle. Partitioning into 64 inner and 128 outer shuttles implies binary addressing via either 6-bit inner and 7-bit outer, or two generic 8-bit binary address fields (the exclusion of the "same shuttle-box" might imply generic addressing).

Ludgate 1909 states "it is important to remember that it is a function of the formula-paper to select the shuttles to receive the Variables, as well as the shuttles to be operated on, so that (except under certain special circumstances, which arise only in more complicated formulae) any given formula-paper always selects the same shuttles in the same sequence and manner, whatever be the values of the Variables". The text in parentheses implies Ludgate allowed selection of a shuttle other than by a literal operand in the formula paper. For example, indirect addressing could be done quite simply by converting the Mill contents to a shuttle selection with another wheel ("Select" wheel) via the Divide and Log 3-digit addressing mechanism; this would be an important aid to execution of complex algorithms. It must be stressed that this is a possible but not the only interpretation.

Ludgate 1909 further states that the shuttles are removable. Maximum flexibility would imply individual shuttles were removable, or whole storage cylinders were removable. This then implies front-loading, and equally implies the races, Index, Mill and Divide, Log, and any other conceivable cylinders were at the rear of the machine.

In order to ensure reasonably fast rotation to select shuttles, the storage cylinder and shuttle inertias would need to be minimised. Riches proposed rectangular rods which slotted together but their weight would give rise to too much inertia. Instead, the cylinders would need to be skeletal (just slotted cylinders), and the shuttles made of a light and strong material, e.g. spring-steel. Ludgate does not say whether the rods are round or square, which will impinge on how well the rods retain their position.

A possible arrangement might have square rods with retention slots on a regular pitch, representing 0-9. If the shuttle sides were made of spring steel and had internal bumps at the same pitch then they would retain the rods in position as well as allow movement of the rods to change value. Lengthwise wobble of the rods could be avoided by ensuring the rod occupied the full width of the shuttle at all times (which would mean the rods would be almost twice as long as the shuttle), perhaps enhanced by bowing the shuttle sides inwards to apply light pressure on the rods. Side bars or shields extending as far as the maximum rod extents in the shuttle would make it easier to register the shuttles at the destination of each movement and would protect the contents while in transit.

Ludgate 1909 states that shuttles are transferred to the Index via races. Provided shuttles, races and Index are oriented in line, either push/pull or hook/move of shuttles seems feasible.

Selecting a shuttle quickly seems problematic. Storage addresses are binary, so successive approximation might be used, perhaps with differentials to successively phase gear ratios, or epicyclic gears [47] to successively approximate addresses. An obvious option, Gray code, was used in Baudot codes c.1878 [48], but not officially invented until 1947 [49]. Alternative simpler approaches using rebound springs and dampers might prove more convincing, as may non-binary selection mechanisms that respond to a binary address.

29

Figure 24 and Figure 25 show an early 3D rendition of a possible re-imagined Ludgate Analytical Machine's storage and Index, correctly dimensioned. From this it can be seen that the cylinders are quite skeletal, hence probably quite light, which will help reduce rotation and positioning inertias, and thereby reduce the necessary control forces.

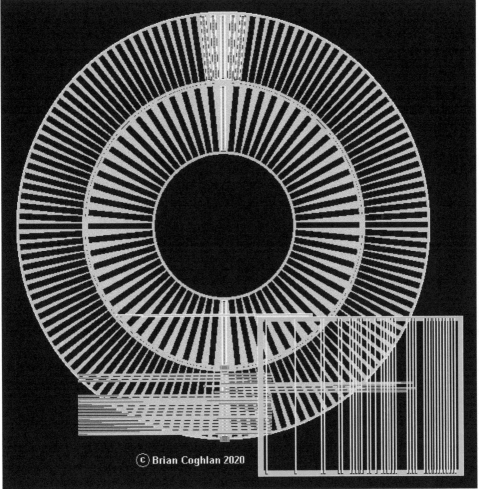

Figure 24 *Early 3D rendition of a possible re-imagined Ludgate's storage and Index*
Top view showing alignment of storage cylinders and shuttles with Index with slides
Showing '813200' multiplied by '9247' as per Engineering [32]
Image reproduced courtesy The John Gabriel Byrne Computer Science Collection

In the design shown in Figure 24, the storage address origin (address '0') is at the top centre, with clockwise increasing addresses. Outer shuttles are shown at outer storage addresses 0, 1, 2, 126, 127, while an inner shuttle is shown at inner storage address 0. In addition, Figure 25 shows that an inner shuttle (holding multiplier '9247') has been withdrawn from inner address 32 along the inner race, and an outer shuttle (holding multiplicand '813200') has been withdrawn from outer address 64 along the outer race. The Index is shown having iterated to the digit '9' of the multiplier.

30

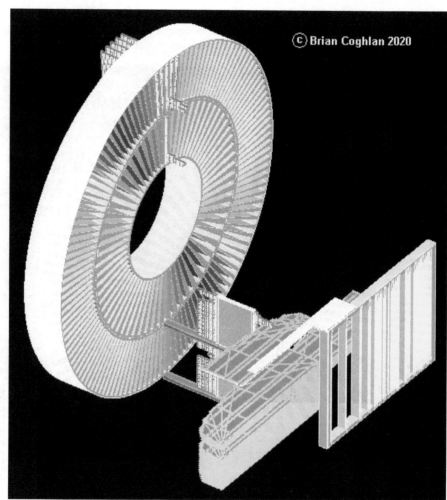

Figure 25 *Early 3D rendition of a possible re-imagined Ludgate's storage and Index*
Three-quarter view showing inner and outer shuttles withdrawn along races
Showing '813200' multiplied by '9247' as per Engineering [32]
Image reproduced courtesy The John Gabriel Byrne Computer Science Collection

IV.10. Ludgate's controlpath

Ludgate 1909 and Engineering [32] are very reticent about the controlpath. Firstly, the former does state "most of the movements … are derived from a set of cams placed on a common shaft parallel to the driving shaft", but says very little else about the mechanism. Secondly and thirdly, and very fortunately, inferences <u>can</u> be made about the core instruction set and preemption.

IV.11. Ludgate's instruction set

What was the format of the basic arithmetic instructions? Ludgate 1909 stated "Each row of perforations across the formula-paper directs the machine in some definite step in the process of calculation—such as, for instance, a complete multiplication, including the selection of the numbers to be multiplied together." Hence each row of the formula tape specifies one instruction, and the arithmetic instruction format must contain an opcode, either a 6-bit inner and 7-bit outer shuttle source operand address, or two 8-bit generic source operand addresses. It is not clear how many arithmetic instructions there were, except for multiply-accumulate (MAC) and in all likelihood multiply (MUL) without accumulation (i.e. pre-clearing Mill). These should be adequate as Ludgate 1909 states that addition is multiply by unity, e.g. MUL(x, 1) and subtraction is, e.g. MUL(x, -1), so the programmer can specify the second operand x as needed. Shifting left could be done with MUL(x, 10^N) and shifting right with

31

MUL(x, 10^{-N}). This makes sense as mechanically only one arithmetic operation (multiply-accumulate) is implemented, and pre-clearing the accumulator would be very simple. However, it may be that Ludgate defined ADD and SUB variants of MUL and MAC that forced just one inner slide iteration rather than a full multiply, with optimizations to skip zeroes.

Lemma 14: the basic arithmetic instruction format contains an opcode.
Lemma 15: the basic arithmetic instruction format contains <u>at minimum</u> a 6-bit inner shuttle source operand address.
Lemma 16: the basic arithmetic instruction format contains <u>at minimum</u> a 7-bit outer shuttle source operand address.

What was the format of the extended arithmetic instructions like division, logarithm, a^n, etc? While the logarithm instruction could act on Mill contents, that could not be "indefinitely extended" to a^n and other such functions. Furthermore, Ludgate 1909 states "When the arrangement of perforations on the formula-paper indicates that division is to be performed, and the Variables which are to constitute divisor and dividend", so the division instruction must have two operands. In other cases the instruction must have at minimum an opcode, but perhaps nothing more. At minimum there would be DIV and LOG opcodes. Ludgate 1909 goes on: "the formula-paper then allows the dividing cylinder to usurp its functions until that cylinder has caused the machine to complete the division", and states the dividing cylinder has perforations like the formula-paper, i.e. the Dividing cylinder is not the same as the Divider cylinder that has holes of depth proportional to reciprocal values. Ludgate 1909 further says "This system of auxiliary cylinders and tables for special formulae may be indefinitely extended."

Lemma 17: the extended arithmetic instruction format contains at minimum an opcode and up to two source operand addresses.

Do the arithmetic instructions include a result address? Ludgate 1909 stated "Each row of perforations … directs the machine in some definite step … for instance, a complete multiplication, including the selection of the numbers to be multiplied together." There is no mention of result addresses, multiply only includes source operands, not destination operands. Ludgate 1909 also stated: "until the whole product ab is found. The shuttles are afterwards replaced in the shuttle-boxes, the latter being then rotated until the second shuttles of both boxes are opposite to the shuttle-race. These shuttles are brought to the index, as in the former case, and the product of their Variables (21893 x 823) is obtained, which, being added to the previous product (that product having been purposely retained in the mill)". Hence multiply results are retained in the Mill. This is consistent with all typical accumulator architectures. Hence arithmetic instructions do not include a result destination address.

Lemma 18: the arithmetic instruction format does not contain a result destination address.

Section IV.5 explains the necessity for decimal-point alignment, most easily solved via a separate alignment instruction.

Lemma 19: there must be a separate alignment instruction.

Was there a separate store instruction? In an accumulator architecture there is little to be gained by integrating result addresses into arithmetic instructions, and neither Ludgate 1909 nor Engineering [32] mention this, hence there must be a separate instruction (STO) to store the Mill contents.

Lemma 20: there must be a separate store instruction.

How many result operands did the store instruction have? Ludgate 1909 stated "or to two shuttles simultaneously, provided that they do not belong to the same shuttle-box" indicating there must be two result operand addresses. That they must not be at the same address might indicate that that a generic 8-bit address is used for both operands.

Lemma 21: the store instruction format must contain two result operand addresses.

What was the format of the store instructions? From Lemma 20 and Lemma 21, the store instruction format must contain an opcode, and inner and outer shuttle result destination addresses consistent with those for arithmetic instructions. It is possible this should allow separate variants to select Divide, Log, and any other conceivable cylinder outputs, for example for a generic 8-bit address A, when $A[7..6]=11$ then individual bits of $A[5..0]$ might select outputs of other variants; alternatively extra opcodes could select those.

Lemma 22: the store instruction format contains an opcode, possibly allowing variants.
Lemma 23: the store instruction format contains <u>at minimum</u> a 6-bit inner shuttle destination address.
Lemma 24: the store instruction format contains <u>at minimum</u> a 7-bit outer shuttle destination address.

What was the format of the conditional instructions? Ludgate 1909 stated "It can also "feel" for particular events in the progress of its work—such, for instance, as a change of sign in the value of a function, or its approach to zero or infinity; and it can make any pre-arranged change in its procedure, when any such event occurs." Hence a conditional instruction format must contain an opcode (e.g. BN | BZ | BV) and a target that associates to a program location (but Ludgate did not discuss location, nor form of addressing). McQuillan [38] makes the interesting suggestion of skip markers (as in old printers) for targets.

Lemma 25: the conditional instruction format contains an opcode (e.g. BN | BZ | BV).
Lemma 26: the conditional instruction format contains a target.
Lemma 27: the conditional instruction target associates to a program location.

Brian Coghlan, Published online, 2022. © 2022, The John Gabriel Byrne Computer Science Collection

Percy Ludgate's Analytical Machine

Very speculatively, the instruction set format could be like that in Table 6 (assuming generic 8-bit shuttle addresses). McQuillan [38] and Riches [42] offer equally speculative but quite different instruction set formats, which highlights the complete absence of any guidance from Ludgate in this regard.

26	25	24	23	22	21	20	19	18	17	16	15	14	13	12	11	10	9	8	7	6	5	4	3	2	1	0
MAC											Shuttle 1 (multiplicand)								Shuttle 2 (multiplier)							
MUL											Shuttle 1 (multiplicand)								Shuttle 2 (multiplier)							
DIV											Shuttle 1 (dividend)								Shuttle 2 (divisor)							
LOG											-								-							
ALN											-								decimal-point position							
STO											Shuttle 1 (destination 1)								Shuttle 2 (destination 2)							
BN											-								target							
BV											-								target							
BZ											-								target							
READ											Shuttle 1 (destination 1)								-							
WRITE											Shuttle 1 (source 1)								-							
STO_DP											-								-							
STO_A											-								-							
HALT											-								-							

Table 6 *Speculative instruction set for Ludgate's analytical machine,
where for addition (ADD) and subtraction (SUB) the programmer must use MAC or MUL with the multiplier = +1 and -1 respectively
READ, WRITE, STO_DP, STO_A are I/O operations and to support preemption*

IV.12. Ludgate's preemption

Does Ludgate's machine support preemption? Ludgate 1909 stated: "Among other powers with which the machine is endowed is that of changing from one formula to another as desired … work of the tabulation can be suspended at any time to allow of the determination by it of one or more results of greater importance or urgency." This is preemption.

Lemma 28: the machine supports preemption.

Has his machine the capabilities required for preemption? If there is only one active set of context objects, then some or all must be physically saved and later restored. Ludgate 1909 stated: "has also the power of recording its results by a system of perforations on a sheet of paper, so that when such a number-paper (as it may be called) is replaced in the machine, the latter can 'read' the numbers indicated thereon, and inscribe them in the shuttles". Hence the machine has the capabilities for preemption.

Lemma 29: the machine has the capabilities required for preemption.

Is this manual or automatic preemption? In its most primitive form, the user must halt execution and run a SUSPEND program that records all the shuttle and Mill contents to a backup number-paper, then the user must mark the formula tape position P with pencil or sticker. To resume, the user must run a RESUME program that reads all the shuttle and Mill contents from the backup number-paper, then the user must place the formula tape at position P and resume execution. Hence at a minimum this is manual preemption just like for early electronic computers.

Lemma 30: at minimum, preemption is manual just like for early electronic computers.

IV.13. Ludgate's input / output

Ludgate 1909 makes it clear his input included two keyboards, and either two perforated paper readers or a single shared reader. For output there are again two perforated paper punches or a single shared punch, and perhaps a printer. With two exceptions, in the absence of new information it does not appear possible to infer any further I/O details, not even whether data I/O interfaced to the datapath and instruction I/O to the controlpath as might be expected. One exception is that Ludgate 1909 stated "Each row of perforations across the formula-paper directs the machine in some definite step in the process of calculation", so each row of the formula tape specifies one instruction. The other exception is that Ludgate 1909 states "The machine prints all results, and, if required, the data, and any noteworthy values which may transpire during the calculation", and then "It also has the power of recording its results by a system of perforations on a sheet of paper"; the fact he discusses one after the other does imply a separate printer of results legible by humans.

33

IV.14. Ludgate's machine performance

For this initial performance analysis, let us speculatively assume Proposition 4 and Proposition 5 are true, i.e. synchronous ripple carry propagation is overlapped with movements of the "carriage near the index", and is essentially invisible. Ludgate 1909 does not mention the time taken by other mechanical movements, so one can also assume those are excluded from his operational timings. Given these assumptions, the Index and Mill mechanisms would <u>effectively</u> conform to Ludgate's statement that the "carrying is practically in complete mechanical independence of the adding process, so that the two movements proceed simultaneously", and would adhere to his statement that "the sum of *m* numbers of *n* figures would take (9*m* + *n*) units of time", where "*n*" represents a final ripple carry at the end of a sequence of *m* additions. Similarly for multiply, where he states the 20-digit multiply-accumulate would involve *m* = 40 additions and would take ((9 x 40) + 20) = 380 time-units.

Let us now consider the timing when mechanical movements are <u>not</u> ignored.

Ludgate states the multiply instruction takes 10 seconds. Firstly there would be access to shuttles from storage and movement along races. Then there would be outer slide movements with movement of the outer frame up and over the inner frame, and 20 inner frame movements. For each inner frame movement there would also be an inner slide movement, a carriage movement to units (1 time-unit), a units add (9 time-units), a carriage movement to tens (1 time-unit), a tens add (9 time-units), and an inner slide retraction. Finally there would be ripple carry propagation of 20 time-units, and decimal-point alignment (although the latter might be statically set). Restoration of shuttles to storage, and retraction of the outer frame and its slides may all be overlapped with the final carry propagation, and so are ignored below. Therefore the time taken would be:

$$T_{storage\_access} + T_{race\_motion} + T_{outer\_slide} + T_{up\_and\_over} + 20*2*T_{inner\_slide} + 20*2*T_{add} + 20*2*T_{carriage} + T_{ripple\_carry} + T_{align}$$
$$= T_{storage\_access} + T_{race\_motion} + T_{outer\_slide} + T_{up\_and\_over} + 20*2*T_{inner\_slide} + 20*2*9*T_{time-unit} + 20*2*T_{time-unit} + 20*T_{time\_unit} + T_{align}$$
$$= T_{storage\_access} + T_{race\_motion} + T_{outer\_slide} + T_{up\_and\_over} + 40*T_{inner\_slide} + 420*T_{time\_unit} + T_{align}$$

Ludgate does not state the duration of a time-unit, only that it is "the period required to move the figure-wheel through $\frac{1}{10}$ revolution". He suggests a "small motor" rotating at 3 revolutions per second, probably an electric (not steam or petrol) motor. Singer produced their first sewing machine with an attached electric motor in 1889. In 1908 when his paper was submitted, Dublin already had 50Hz AC mains electricity (first installed in 1892, substantially upgraded in 1903) [50]. So Ludgate's motor would be 50Hz, synchronous or asynchronous, but likely synchronous as it must rotate at an "approximately uniform rate of rotation". Some sensible possibilities are:

(1) A time-unit is $\frac{1}{10}$ of a motor revolution, i.e. $\frac{1}{30}$ seconds, so then the core arithmetic actions take:

$$(40*T_{inner\_slide} + 420*T_{time-unit}) > 14 \text{ sec}$$

which is longer than the stated 10 seconds taken by multiply. Therefore a time-unit must be much shorter than $\frac{1}{30}$ seconds.

(2) A time-unit is one cycle of 50Hz, i.e. $\frac{1}{50}$ seconds, so then the core arithmetic actions take:

$$(40*T_{inner\_slide} + 420*T_{time-unit}) > 8.4 \text{ sec}$$

But 50Hz does not evenly divide to 3 revs/sec. It is approx 50Hz divided by 17 (a prime number, which might have appealed to Ludgate) yielding 2.94 revs/sec, or alternatively divided by 16 yielding 3.125 revs/sec. In addition one iteration of the core arithmetic actions (2 adds + 2 carriage movements) takes:

$$(2*T_{inner\_slide} + 20*T_{time-unit}) > 0.4 \text{ sec}$$
$$\text{i.e. iteration proceeds at} < 2.5\text{Hz}$$

which does not match well with a 3 revs/sec motor.

(3) One iteration of the core arithmetic actions takes one revolution of the motor at 3 revs/sec:

$$(2*T_{inner\_slide} + 20*T_{time-unit}) = \tfrac{1}{3} \text{ sec}$$
$$\text{i.e. } 20*T_{time-unit} = \tfrac{1}{3} - 2*T_{inner\_slide} \quad\dots\dots\dots\dots\dots\dots\dots\dots\dots\dots\dots\dots\text{ [E]}$$
$$\text{i.e. } T_{time-unit} < \tfrac{1}{60} \text{ sec} \quad\dots\dots\dots\dots\dots\dots\dots\dots\dots\dots\dots\dots\dots\text{ [F]}$$

and therefore the core arithmetic actions (20 iterations + ripple carry) take:

$$(20*\tfrac{1}{3} + 20*T_{time\_unit}) = (6.67 + 20*T_{time\_unit}) \text{ sec} < 7 \text{ sec}$$

and hence a useful deduction, that we can use further below, is:

$$T_{storage\_access} + T_{race\_motion} + T_{outer\_slide} + T_{up\_and\_over} < 3 \text{ sec} \quad\dots\dots\dots\dots\dots\dots\dots\dots\dots\text{ [G]}$$

Option (3) seems the most likely in the absence of other information.

Let us assume option (3) above for an analysis of the addition or subtraction operations, which Ludgate 1909 states take 3 seconds. Ludgate's addition is multiply by +1, and subtraction is multiply by -1 via reversed wheel direction. In both cases partial product Tens will be zero, so no Tens iterations will be done.

When mechanical movements <u>are</u> ignored, "*m*" 1-digit adds will take (9 + *n*) ticks as Ludgate states, so addition or subtraction will take (9 + 20) = 29 time-units.

Percy Ludgate's Analytical Machine

When mechanical movements are <u>not</u> ignored, firstly there would be access to shuttles from storage and movement along races. Then there would be outer slide and frame movements, but no inner frame iterations. There would only be one inner slide movement, a carriage movement to units (1 time-unit), a units add (9 time-units), a carriage movement to tens (1 time-unit), <u>no</u> tens add, and an outer slide retraction. Finally there would be ripple carry propagation of 20 time-units, and decimal-point alignment (again, the latter might be statically set). Restoration of shuttles to storage, and retraction of the outer frame and its slides may all be overlapped with the final carry propagation, and so are ignored below. Therefore the time taken would be:

$$T_{storage\_access} + T_{race\_motion} + T_{outer\_slide} + T_{up\_and\_over} + 2*T_{inner\_slide} + T_{add} + 2*T_{carriage} + T_{ripple\_carry} + T_{align}$$
$$= T_{storage\_access} + T_{race\_motion} + T_{outer\_slide} + T_{up\_and\_over} + 2*T_{inner\_slide} + 31*T_{time\_unit} + T_{align}$$

Applying the inequaities [F] and [G] above, the time taken for addition would be:

$$< 3 + 31*\tfrac{1}{60} \text{ sec, i.e.} < 3.5 \text{ sec}$$

This is a reasonable indication that a time-unit must be $\frac{1}{60}$ seconds or less.

Lemma 31: a time-unit must be approximately $\frac{1}{60}$ seconds or less.
Lemma 32: $T_{storage\_access} + T_{race\_motion} + T_{outer\_slide} + T_{up\_and\_over} < 3$ sec

But in fact the inner slide is firmly mounted to the lower frame, with far more inertia than outer slides, and hence the time taken to move the inner slide cannot be ignored. From equation [E] above:

$$T_{time-unit} = \tfrac{1}{60} - \tfrac{1}{10}*T_{inner\_slide}$$

So the time-unit must be much less than $\frac{1}{60}$ seconds. By way of example, for a time-unit of $\frac{1}{100}$ seconds, the addition time is:

$$< 3 + 31*\tfrac{1}{100} \text{ sec, i.e.} < 3.3 \text{ sec}$$

All the above underrates the time taken shifting the inner frame, exacerbated by its reciprocating nature, and also speculatively assumes Proposition 4 and Proposition 5 are true. Nonetheless, the claims made would indicate the machine was fast for 1909. Ludgate stated that addition and subtraction would each take 3 seconds, multiplication 10 seconds, division 90 seconds, logarithms 120 seconds, and a^n would take 210 seconds..

To compare with a basic (un-optimized) addition-based machine let us ignore mechanical movements and assume it handles the same 20 digits as Ludgate's machine, and adopt his definition of a time-unit as the time to rotate a Mill wheel by a $\frac{1}{10}$ turn. For Ludgate's machine to sum m numbers of $n = 20$ figures would take $(9m + n) = (9m + 20)$ time-units, while for a basic addition-based machine it would take $m(9 + n) = 29m$ time-units, i.e. Ludgate's machine would be faster for $m > 1$. For Ludgate's machine to multiply two numbers of 20 figures would take 380 time-units, while a basic addition-based machine would take $N(9 + n) = 29N$ time-units, best when N is the smallest of the two numbers, i.e. Ludgate's machine would be faster for $N > 14$. Since Ludgate's machine multiplies in 10 seconds but divides in 90 seconds, i.e. division is 9 times slower, then by extrapolation division would take approximately 3420 time-units, while for a basic addition-based machine it would take $29N$ time-units, best when N is the largest of the two numbers, i.e. Ludgate's machine would be faster for $N > 118$.

Of course, an addition-based machine could be optimized, for example, Babbage's later analytic engine designs could accumulate a 40-digit number in 3 seconds and multiply 20 digits by 40 digits in 120 seconds [51].

IV.15. Was Ludgate's machine Turing-complete?

An 'analytical machine' is equivalent to a general-purpose computer, and (ignoring physical limits) in theory can be programmed to solve any solvable problem. It is "Turing-complete", or more specifically it is a Universal Turing Machine, i.e. one that can emulate any other Turing Machine, a term invented to reflect Alan Turing's contribution to the theory of computing. The term 'analytical machine' should not be confused with the names of the first two of the only three mechanical designs before the electronic computer era: in 1843 Charles Babbage's "Analytical Engine", then in 1909 Percy Ludgate's very different "Analytical Machine" (the third was in 1936 Zuse's "V1", later renamed "Z1"). From c.1914 electromechanical designs began, and c.1937 electronic designs began [11].

None of these three mechanical computers were "natively" Turing-complete, instead they could be programmed to solve a (substantial) subset of all solvable problems, i.e. they were multi-purpose rather than general-purpose machines. For a discussion of whether and how Ludgate's machine could be rendered Turing-complete, see Appendix IV.

35

IV.16. Re-imagining Ludgate's machine

Drawing substantially on previous efforts by Riches [42], and McQuillan [37][38], and on the above, it may be possible to at least re-imagine parts of Ludgate's machine.

Objective 2: re-imagine parts of Ludgate's machine.

The aim might be to enable this in several phases:

(1) Datapath Simulation: this is work in progress using Python. The intention is to explore mechanical options, with unit tests for correctness, see Figure 26.

(2) Hybrid CAD Simulation: a 3-D CAD model of a mechanical datapath plus algorithmic controlpath. The intention could be that this should yield online simulations that would create downloadable video files of execution of web user's programs.

(3) Hybrid Engine: a 3D-printed mechanical datapath, with an electronic controlpath processor like a Beaglebone, Arduino or Raspberry Pi, with the controlpath, program and input/output web-enabled and/or on an Android or IOS App.

(4) Full CAD Simulation: a 3-D CAD model of a fully mechanical datapath plus controlpath, taking the results of soft testing controlpath proposals on hybrid implementations. Again the intention could be to yield online simulations and executions.

(5) Pure Engine: fully mechanical engine, manufactured from the 3-D design files either by 3-D printing or using numerically-controlled machine tools. Small programs (e.g. Bernoulli) could be on "real" formula tapes for demonstrations.

The premise might be that mechanical construction of the datapath is easiest, followed by storage, then the controlpath (sequencer), and finally input/output. It is expected it would take several years to fruition, and be very speculative.

```
The following options are supported though there is not command interface yet.

"size"      "full" or "test"
            test has only 5 digits in a shuttle and 16+8 shuttles in the culinders
            test also defaults to immediate execution rather than cycle delays
"mill"      "single" or "double" width of mill and how it works
"extra"     extra digits at the end of the mill to get last digit meaningful
"shuttle"   "ordinal" or "decimal" format of the dgits in the shuttles
"delay"     time for a main shaft rotation default 1/3 sec or 0 for test

Operations: Numbers are represented as digits followed by a sign
            Cylinders are 0 inner and 1 outer, addr 0 to boxes in cylinder - 1

verbose level                   set tracing verbose level,
                                0 - none, 1 - commands,
                                2 - the workings, 3 - debug
print message                   print string as annotation.
time                            print current time and time since last call
set cyl addr value              set cylinder shuttle to a value
                                value is digits followed by a sign
show cyl addr                   display the value in the shuttle
set_dp dest, src1, src2         set decimal places in op destination and sources
load cyl addr                   load the mill with dest src1 decimal points
load_neg cyl addr               load the mill with negative of operand
store cyl addr                  store value from the mill
store2 addr1 addr2              store mill value in outer and inner cylinders
add cyl addr                    accumulate value in the mill
sub cyl addr                    subtract value from accumulator
mul inner_adr outer_adr         multiply sources from the two shuttles
mul_add inner_adr outer_adr     multiply and accumulate
mul_sub inner_adr outer_adr     subtract multiply from accumulator

Shuttles 5 digits + sign, format decimal          (C) David McQuillan 2020
Cylinders Outer 16 boxes, Inner 8 boxes
Mill 6 digits (1 extra), format single
Delay 0
Verbose set to 1
 * Op.set 0 1 98+
 * Op.set 1 5 37+
Verbose set to 2
 * Op.mul inner[1] outer[5]
   Address Outer box 5
   Address Inner box 1
   Move shuttle to Frame from Outer
   Move shuttle to Slide from Inner
   Frame 00037+ ordinals 99948  index [50, 50, 50, 7, 33]
   Slide 00098+ ordinals 99963 index [50, 50, 50, 14, 3]
   Clear the mill
   set_mul_sign_position
   Set mill position -5 to add partial product from frame
   Partial products 50 + [50, 50, 50, 7, 33] = [100, 100, 100, 57, 83]
       = decimal [0, 0, 0, 0, 0]
```

Figure 26 *Help text from an early version of a Python simulator of the datapath of Ludgate's Analytical Machine*
Image reproduced courtesy David McQuillan

36

V. CONCLUDING REMARKS

The Christmas 2019 the discoveries of the articles and especially the diagram in "The English Mechanic", then the discovery of Engineering [32], have led to intensive efforts to deduce hidden details of Ludgate's analytical machine. From the latter efforts, thus far a surprising amount of understanding has emerged, construction detail has been uncovered, some useful dimensional, geometric and timing inequalities established, and an effort has been made to codify these new facts. Work has begun on both simulating the machine and "re-imagining" it with modern engineering software.

ACKNOWLEDGMENTS

The author would like to thank Professor Brian Randell and David McQuillan (for their engagement, support, and robust debate about the many speculative issues that arise in relation to Ludgate's machine), Professor John Tucker, University College Swansea, UK (for access to the Riches report on Ludgate), Royal Dublin Society Library (for access to records), Ralf Buelow of Heinz Nixdorf MuseumsForum and Eric Hutton (for discovery of the articles in "The English Mechanic" and permission to publish the images), and Jade Ward of the University of Leeds Library (for discovery of the article of Engineering [32] and permission to publish the images). Finally my thanks for the support of the School of Computer Science and Statistics, Trinity College Dublin, for this work and for The John Gabriel Byrne Computer Science Collection [8].

APPENDIX I

FROM: ENGINEERING, AUG. 20, 1909, PAGES 256-257 [32]

OPTICAL CHARACTER RECOGNITION COURTESY DAVID MCQUILLAN

A PROPOSED ANALYTICAL MACHINE

By name, at any rate, Babbage's famous analytical engine is known to all. It was intended to be a machine for the arithmetical solution of all problems in mathematical physics. Such solutions are generally, perhaps always, feasible, but in most cases when the computations have to be effected by direct human agency, they are so extremely tedious as to be practically, if not theoretically, impossible. Every operation in arithmetic can be reduced to addition, subtraction, multiplication, and division, and, indeed, the two latter operations can be regarded as mere extensions of the two former. The analytical engine was a machine by which these four operations could be performed in any desired sequence ; moreover, a number of partial operations could be combined, and the final results automatically tabulated for any required values of the variable. As is well known, though many years' labour was spent on the machine, it was never even partially completed, Mr. Babbage's scheme being far too ambitious for a first effort. He wished, indeed, to tabulate values to 50 significant figures, thus enormously complicating the mechanism and augmenting the cost of the experiment. In a paper read not long ago before the Royal Dublin Society, Mr. Percy E. Ludgate has revived again the idea of constructing such a machine: As proposed by him, the machine differs from that of Babbage in some fundamental details, though, as in its predecessor, Jacquard cards will be used to control the sequence of operations. Thus if, for instance, a number of values of the series

$$y = x - \frac{x^2}{2^2} + \frac{x^3}{2^2.3^2} - \frac{x^4}{2^2.3^2.4^2} + \&c.$$

were required, the appropriate card would be placed in the machine, which would then, for different values of x, calculate each term of the series, add all the positive terms together, subtract from this sum all the negative, and print the result. For a different series a different card would be used.

In Babbage's engine it was proposed to effect multiplication by successive additions, and divisions by successive subtractions, just as is now done in the case of the ordinary arithmometer. Mr. Ludgate, in his engine, proposes to effect these operations on entirely different principles. Multiplication is effected by a series of index numbers analogous to logarithms;

The arrangement is shown diagrammatically in Fig. 1 *[see Figure 29]*. Here the number 813,200 is to be multiplied by 9247. The arrow under 8 represents a slide, which to denote 8 is set at $\frac{3}{8}$ in. above the zero or starting line. The slide representing 1 lies

on the starting line, whilst that representing 3 stands $\frac{7}{8}$ in. above this line, and that corresponding to the number two $\frac{1}{8}$ in. above.

On the other hand, the slides representing zero are set 50 eighths above the starting line. The number of units above the starting line corresponding to each digit of the multiplicand are known as index numbers, and a complete table of these has been drawn up by Mr. Ludgate. All the slides aforementioned are mounted in a frame, and to multiply by 9, this frame is moved up over another frame divided with another series of index numbers. Thus, as shown, the distance between the lower frame and the starting line is such that the top of this lower frame lies on the index number corresponding to 9; that is, 14 eighths below the starting line. The lower end of the No. 8 slide, represented by the black circle, rests then, it will be seen, on a line marked "72," which is the product of 8 and 9. The digits 7 and 2 appear accordingly on the register below. Similarly, the tail of the No. 1 slide rests on the No. 9 line, that of the No. 3 slide on the No. 27 line, and that of the No. 2 slide on 18, corresponding to the partial products 9 × 1; 9 × 3 ; and 9 × 2. The tails of the zero slides rest on no line in the lower frame, and hence zero is registered for these. All these partial products are registered in the mill below, as indicated. In a final operation these partial products are added together as indicated, giving 7,318,800. If now the frame is moved to the index number below the starting line marked "2," it will be found, on trial with a piece of tracing paper, that the tail of the No. 8 slide rests now on the line marked "16" — i.e., 8 × 2. That of the No. 1 slide on the index-line marked "2," that of slide 3 on the line marked "15," and that of the No. 2 slide on the line marked "10." These partial products will then appear on the mill and be added together, giving the result of the multiplication of 813,200 by 2. The process is repeated for the remaining figures of the multiplier, and the whole added together so as to give the product of 813,200 × 9247. Mr. Ludgate proposes to give such products to twenty significant figures, the time required being, he states, about 10 seconds.

To divide one number by another he proceeds in a different fashion. He notes that the expression $\frac{p}{q}$, where p and q are any two numbers, can always be expressed in the form—

$$\frac{p}{q} = \frac{A\,p}{1+x},$$

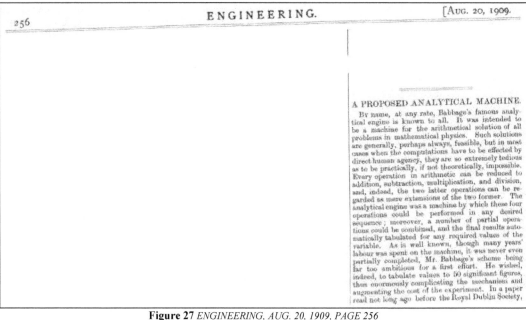

38

where x is a small quantity, and A is the reciprocal of some number between 100 and 999.

The above expression can also obviously be written

$$\frac{p}{q} = A\,p\,(1 - x + x^2 - x^3 + x^4 - x^5, +\&c.),$$

the series being very rapidly convergent, the first eleven terms give the value of $\frac{1}{1+x}$ correct to at least twenty figures.

He proposes to perform division, therefore, by making the machine first calculate the value of this series, after which it will multiply $A\,p$ by the value thus found. As a maximum, he considers that this operation giving the result correct to twenty figures might require $1\frac{1}{2}$ minutes.

Figure 27 *ENGINEERING, AUG. 20, 1909, PAGE 256*
Image reproduced courtesy Jade Ward, University of Leeds Library

Mr. Percy E. Ludgate has revived again the idea of constructing such a machine. As proposed by him, the machine differs from that of Babbage in some fundamental details, though, as in its predecessor, Jacquard cards will be used to control the sequence of operations. Thus if, for instance, a number of values of the series

$$y = x - \frac{x^3}{2^2} + \frac{x^5}{2^2 . 3} - \frac{x^7}{2^2 . 3 . 4^2} + \&c.$$

were required, the appropriate card would be placed in the machine, which would then, for different values of x, calculate each term of the series, add all the positive terms together, subtract from this sum all the negative, and print the result. For a different series a different card would be used.

In Babbage's engine it was proposed to effect multiplication by successive additions, and divisions by successive subtractions, just as is now done in the case of the ordinary arithmometer. Mr. Ludgate, in his engine, proposes to effect these operations on entirely different principles. Multiplication is effected by a series of index numbers analogous to logarithms.

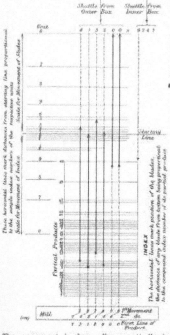

The arrangement is shown diagrammatically in Fig. 1. Here the number 813,200 is to be multiplied by 9247. The arrow under 8 represents a slide, which to denote 8 is set at ⅞ in. above the zero or starting line. The slide representing 1 lies on the starting line, whilst that representing 3 stands ¼ in. above this line, and that corresponding to the number two ½ in. above. On the other hand, the slides representing zero are set 50 eighths above the starting line. The number of units above the starting line corresponding to each digit of the multiplicand are known as index numbers, and a complete table of these has been drawn up by Mr. Ludgate. All the slides aforementioned are mounted in a frame, and to multiply by 9, this frame is moved up over another frame divided with another series of index numbers. Thus, as shown, the distance between the lower frame and the starting line is such that the top of this lower frame lies on the index number corresponding to 9; that is, 14 eighths below the starting line. The lower end of the No. 8 slide, represented by the black circle, rests then, it will be seen, on a line marked "72," which is the product of 8 and 9. The digits 7 and 2 appear accordingly on the register below. Similarly, the tail of the No. 1 slide rests on the No. 9 line, that of the No. 3 slide on the No. 27 line, and that of the No. 2 slide on 18, corresponding to the partial products 9 × 1;

9 × 3; and 9 × 2. The tails of the zero slides rest on no line in the lower frame, and hence zero is registered for these. All these partial products are registered in the mill below, as indicated. In a final operation these partial products are added together as indicated, giving 7,318,800. If now the frame is moved to the index number below the starting line marked "2," it will be found, on trial with a piece of tracing paper, that the tail of the No. 8 slide rests now on the line marked "16"—i.e., 8 × 2. That of the No. 1 slide on the index-line marked "2," that of slide 3 on the line marked "15," and that of the No. 2 slide on the line marked "10." These partial products will then appear on the mill and be added together, giving the result of the multiplication of 813,200 by 2. The process is repeated for the remaining figures of the multiplier, and the whole added together so as to give the product of 813,200 × 9247. Mr. Ludgate proposes to give such products to twenty significant figures, the time required being, he states, about 10 seconds.

To divide one number by another he proceeds in a different fashion. He notes that the expression $\frac{p}{q}$, where p and q are any two numbers, can always be expressed in the form—

$$\frac{p}{q} = \frac{A p}{1 + x},$$

where x is a small quantity, and A is the reciprocal of some number between 100 and 999. The above expression can also obviously be written

$$\frac{p}{q} = A p (1 - x + x^2 - x^3 + x^4 - x^5, + \&c.),$$

the series being very rapidly convergent, the first eleven terms give the value of $\frac{1}{1 + x}$ correct to at least twenty figures.

He proposes to perform division, therefore, by making the machine first calculate the value of this series, after which it will multiply A p by the value thus found. As a maximum, he considers that this operation giving the result correct to twenty figures might require 1½ minutes.

Figure 28 *ENGINEERING, AUG. 20, 1909, PAGE 257*
Image reproduced courtesy Jade Ward, University of Leeds Library

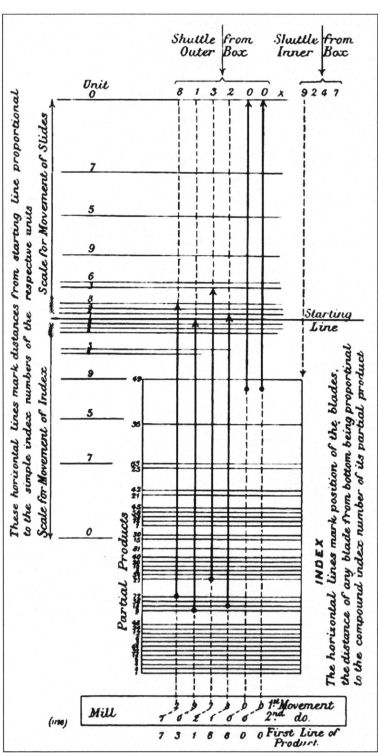

Figure 29 *Diagram from "Engineering", Aug. 20, 1909, Pages 256-257 [32]*
Image reproduced courtesy Jade Ward of University of Leeds Library

41

APPENDIX II

FROM: ENGLISH MECHANIC AND WORLD OF SCIENCE, NO. 2302, MAY 7, 1909, PAGE 322 [30]

OPTICAL CHARACTER RECOGNITION COURTESY DAVID MCQUILLAN

SCIENTIFIC NEWS

No. 9 of the Scientific Proceedings of the Royal Dublin Society (Williams and Negate, 14, Henrietta-street, WIC., 6d.). contains an interesting paper by Mr. Percy E. Ludgate describing a proposed new analytical machine designed by him with the object of designing machinery capable of performing calculations, however intricate or laborious, without the immediate guidance of the human intellect. Mr. Ludgate says:—

"Babbage's Jacquard system and mine differ considerably ; for, while Babbage designed two sets of cards—one set to govern the operations, and the other set to select the numbers to be operated on—I use one sheet or roll of perforated paper (which, in principle, exactly corresponds to a set of Jacquard cards) to perform both these functions in the order and manner necessary to solve the formula to which the particular paper is assigned. . . .

In my machine each variable is stored in a separate shuttle, the individual figures of the variable being represented by the relative positions of protruding metal rods or "type," which each shuttle carries. There is one of these rods for every figure of the variable, and one to indicate the sign of the variable. Each rod protrudes a distance of from 1 to 10 units, according to the figure or sign which it is at the time representing. The shuttles are stored in two co-axial cylindrical shuttle-boxes, which are divided for the purpose into compartments parallel to their axis. The present design of the machine provides for the storage of 192 variables of twenty figures each ; but both the number of variables and the number of figures in each variable may, if desired, be greatly increased. It may be observed, too, that the shuttles are quite independent of the machine, so that new shuttles, representing new variables, can be introduced at any time."

Figure 30 *ENGLISH MECHANIC AND WORLD OF SCIENCE, NO. 2302, MAY 7, 1909, PAGE 322*
Image reproduced courtesy Ralf Buelow of Heinz Nixdorf MuseumsForum, and Eric Hutton

APPENDIX III

FROM: ENGLISH MECHANIC AND WORLD OF SCIENCE, NO. 2319, SEPT 3, 1909, PAGE 111 [31]

OPTICAL CHARACTER RECOGNITION COURTESY DAVID MCQUILLAN

A PROPOSED ANALYTICAL MACHINE

Mr. Percy E. Ludgate has revived again the idea of constructing such a machine. As proposed by him, the machine differs from that of Babbage in some fundamental details, though, as in its predecessor, Jacquard cards will be used to control the sequence of operations. Thus if, for instance, a number of values of the series

$$y = x - \frac{x^2}{2^2} + \frac{x^3}{2^2.3^2} - \frac{x^4}{2^2.3^2.4^2} + \&c.$$

were required, the appropriate card would be placed in the machine, which would then, for different values of x, calculate each term of the series, add all the positive terms together, subtract from this sum all the negative, and print the result. For a different series a different card would be used.

In Babbage's engine it was proposed to effect multiplication by successive additions, and divisions by successive subtractions, just as is now done in the case of the ordinary arithmometer. Mr. Ludgate, in his engine, proposes to effect these operations on entirely different principles. Multiplication is affected by a series of index numbers analogous to logarithms. Say the number 813,200 is to be multiplied by 9,247. A slide to denote 8 is set at $\frac{3}{8}$ in. above the zero- or starting-line. The slide representing 1 lies on the starting-line, whilst that representing 3 stands $\frac{7}{8}$ in. above this line, and that corresponding to the number two $\frac{1}{8}$ in. above. On the other hand, the slides representing zero are set 50 eighths above the starting-line. The number of units above the starting-line corresponding to each digit of the multiplicand are known as index numbers, and a complete table of these has been drawn up by Mr. Ludgate. All the slides aforementioned are mounted in a frame, and to multiply by 9, this frame is moved up over another frame divided with another series of index numbers. Thus, as shown, the distance between the lower frame and the starting-line is such that the top of this lower frame lies on the index number corresponding to 9; that is, 14 eighths below the starting-line. The lower end of the No. 8 slide, represented by the black circle rests, then, it will be seen, on a line marked "72," which is the product of 8 and 9. The digits 7 and 2 appear accordingly on the register below. Similarly, the tail of the No. 1 slide rests on the No. 9 line, that of the No. 3 slide on the No. 27 line, and that of the No. 2 slide on 18, corresponding to the partial products 9 × 1; 9 × 3; and 9 × 2. The tails of the zero slides rest on no line in the lower frame, and hence zero is registered for these. All these partial products are registered in the mill below, as indicated. In a final operation, these partial products are added together as indicated, giving 7,318,800. If now the frame is moved to the index number below the starting-line marked "2," it will be found, on trial with a piece of tracing-paper, that the tail of the No. 8 slide rests now on the line marked "16 "—i.e., 8 × 2. That of the No. 1 slide on the index-line marked "2," that of slide 3 on the line marked "15," and that of the No. 2 slide on the line marked "10." These partial products will then appear on the mill, and be added together, giving the result of the multiplication of 813,200 by 2. The process is repeated for the remaining figures of the multiplier, and the whole added together so as to give the product of 813,200 × 9,247. Mr. Ludgate proposes to give such products to twenty significant figures, the time required being, he states, about 10 seconds.

To divide one number by another, he proceeds in a different fashion. He notes that the expression $\frac{p}{q}$, where p and q are any two numbers, can always be expressed in the form—

$$\frac{p}{q} = \frac{A p}{1 + x},$$

where x is a small quantity, and A is the reciprocal of some number between 100 and 999.

The above expression can also obviously be written

$$\frac{p}{q} = A p (1 - x + x^2 - x^3 + x^4 - x^5, + \&c.),$$

the series being very rapidly convergent, the first eleven terms give the value of $\frac{1}{1+x}$ correct to at least twenty figures.

He proposes to perform division, therefore, by making the machine first calculate the value of this series, after which it will multiply $A p$ by the value thus found. As a maximum, he considers that this operation, giving the result correct to twenty figures, might require $1\frac{1}{2}$ minutes.—"Engineering."

43

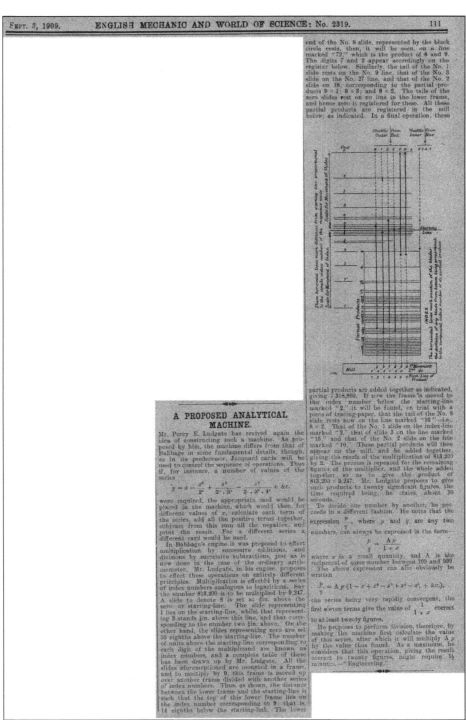

Figure 31 *ENGLISH MECHANIC AND WORLD OF SCIENCE, NO. 2319, SEPT 3, 1909, PAGE 111*
Image reproduced courtesy Ralf Buelow of Heinz Nixdorf MuseumsForum, and Eric Hutton

44

APPENDIX IV

WAS LUDGATE'S MACHINE A GENERAL-PURPOSE COMPUTER?

The term *general-purpose computer* is somewhat controversial. It has been equated to the von Neumann architecture, but this is questionable, although the converse is more assured, that machines using the von Neumann architecture are general-purpose computers. One could expect that a general-purpose computer could execute any algorithm. But an algorithm is a special-purpose Turing Machine, therefore one might conjecture that a general-purpose computer (itself a Turing Machine) should if truly 'general-purpose' be able to execute any other Turing Machine. A Universal Turing Machine can emulate any other Turing Machine, therefore a reasonable pragmatic definition might be that a general-purpose computer is a physical realization of a Universal Turing Machine.

But a Universal Turing Machine stores both code and data on its (infinite) tape, i.e. code and data share the same address space and access mechanisms. This facilitates self-modifying code, and more importantly access to code or data as a result of prior calculation, i.e. indirect access to code or data, which is crucial for many algorithms. Therefore it is furthermore reasonable to conclude that a general-purpose computer must at minimum facilitate indirect access to code or data.

Raul Rojas of the Freie Universitat Berlin, Germany, has explored the programming architecture of Babbage's engine, and states that Babbage's engine did not have the ability to indirectly access memory [55]. But Rojas also speculates how a sequence of numbers could have been added using an addition loop, e.g. indirectly with one single operation card and combinatorial cards.

Similarly from Ludgate 1909 it can be stated that his Analytical Machine did not have the explicit ability to indirectly access shuttles, since there is no explicit discussion of any support for selection of a shuttle based on the result of a prior calculation. But Ludgate does hint (see Section IV.9 on Ludgate's Storage above) that for complex formulae it may have been possible to select a shuttle other than by a literal operand, i.e. indirectly.

Nevertheless neither of Babbage's or Ludgate's machines explicitly facilitate indirect access to code or data, and hence could not be said to <u>natively</u> be general-purpose machines. In this Appendix we explore to what extent Ludgate's Analytical Machine could have <u>emulated</u> a general-purpose computer, and represent that with some speculative "Ludgate assembly language" code.

Raul Rojas also showed how Zuse's c.1941 "Z3" could be rendered Turing-complete [56][57][58][59]. The Z3 is not considered to be natively Turing-complete because it didn't have conditional branching (it was designed to solve aircraft flow equations). However, Rojas shows the Z3 could be made Turing-complete by forming a loop of the film used in the Z3 to hold programs, plus a lot of mental gymnastics. Perhaps the Z1 could also be made Turing-complete in the same way (the two designs have the same origins), but this is not yet known.

For Ludgate's machine a loop could also be made, in this case of the formula paper used in his machine to hold programs. But instead of mental gymnastics, the loop could contain an emulator program that continuously cycled until halted. For simplicity the emulator could simply emulate the Ludgate machine itself, but with two enhancements: (a) to execute from programs in shuttles, and (b) to execute one extra instruction to do dynamic shuttle selection (indirect addressing). That would allow straightforward emulation of a von Neumann machine.

Rojas' paper [58] also states that recursion was implemented in the Z3 by unrolling code to refer to absolute addresses. This is useful in that it shows a precedent, even if Ludgate was unaware of it. Here we suggest emulating indirect access to Ludgate's 192 storage shuttles by in effect unrolling code that would loop until it found the instruction to access the specific desired shuttle. Ludgate's formulae paper is unlimited ('infinite' a la Turing?) so the loop unrolling for 192 shuttles (about 1.5k rows of instructions on the formulae paper) is of minimal consequence (perhaps successive approximation could be used to reduce the time taken to get to the desired instruction).

```
#
# Stored-program emulator for Ludgate's Analytical Machine.
# IN DEVELOPMENT
#
```

45

APPENDIX V

OCCAM CODE FOR LUDGATE'S INDEX AND MILL

The behaviour of Ludgate's Index and Mill is relatively concisely represented above as an algorithmic state machine (ASM) using a pseudo-OCCAM. This Appendix expands on how these mechanisms might actually have supported that by transforming the pseudo-OCCAM into a very much more lengthy form of real OCCAM code.

```
#
# IN DEVELOPMENT
#
```

46

REFERENCES

[1] Percy E.Ludgate, On a Proposed Analytical Machine, Scientific Proceedings of the Royal Dublin Society, Vol.12, No.9, pp.77–91, 28-Apr-1909.
[2] C.V.Boys, A new analytical engine, Nature, Vol.81, pp.14-15, Jul-1909, see elsewhere in the Literature category of this catalog.
[3] Percy E.Ludgate, Automatic Calculating Machines, In "Handbook of the Napier Tercentenary Celebration or modern instruments and methods of calculation", Ed: E.M.Horsburgh, 1914
[4] Lovelace, A., Sketch of the Analytical Engine invented by Charles Babbage, Esq. By L.F. Menabrea, of Turin, Officer of the Military Engineers, with notes by the Translator, Scientific Memoirs, Ed.R.Taylor, Vol.3, pp.666–731, 1843.
[5] Baxandall, D., Calculating machines and instruments: Catalogue of the Collection in the Science Museum, Science Museum, London, 1926.
[6] Brian Randell, "Ludgate's analytical machine of 1909", The Computer Journal, Vol.14, No.3, pp.317-326, 1971.
[7] Brian Randell, "From analytical engine to electronic digital computer: The contributions of Ludgate, Torres, and Bush", Annals of the History of Computing, Vol.4, No.4, IEEE, Oct. 1982.
[8] Trinity College Dublin, The John Gabriel Byrne Computer Science Collection. Available: https://scss.tcd.ie/SCSSTreasuresCatalog/
[9] Trinity College Dublin, Percy E. Ludgate Prize in Computer Science. Available: https://scss.tcd.ie/SCSSTreasuresCatalog/miscellany/TCD-SCSS-X.20121208.002/TCD-SCSS-X.20121208.002.pdf
[10] Brian Coghlan, Brian Randell, Paul Hockie, Trish Gonzalez, David McQuillan, Reddy O'Regan, 'Investigating the work and life of Percy Ludgate', *IEEE Annals of the History of Computing*, Vol.43, No.1 pp.19-37, 2021.. Available in file:
"IEEE-Annals-FINAL-asPublished-Ludgate-20210307-0849" at
https://scss.tcd.ie/SCSSTreasuresCatalog/miscellany/TCD-SCSS-X.20121208.002/
[11] Brian Randell (Ed.), The Origins of Digital Computers: Selected Papers (3rd ed.), pp.580, Springer-Verlag, Heidelberg, 1982.
[12] Sydney Padua, The Thrilling Adventures of Lovelace and Babbage, pp.320, Penguin, 2015.
[13] Gottfried Wilhelm Liebniz, Stepped Reckoner, c.1671, Available: https://en.wikipedia.org/wiki/Stepped_reckoner
[14] Joseph Marie Jacquard, Jacquard Loom, c.1804. Available at: https://en.wikipedia.org/wiki/Joseph_Marie_Jacquard
[15] Plan28, Building Charles Babbage's Analytical Engine. Available at: http://www.plan28.org/
[16] Royal Dublin Society, Minutes of the RDS Scientific Committee for 1908-1909, RDS Library, Dublin.
[17] B. Coghlan, B. Randell, P. Hockie, T. Gonzalez, D. McQuillan, R. O'Regan, "Percy Ludgate (1883-1922), Ireland's First Computer Designer ", *Proceedings of the Royal Irish Academy: Archaeology, Culture, History, Literature*, Volume 121C, 2021.
[18] Adrian Johnstone, Notions and notation: Babbage's language of thought, Royal Holloway. Available at: https://babbage.csle.cs.rhul.ac.uk/. Also at: https://pure.royalholloway.ac.uk/portal/en/projects/notions-and-notation-babbages-language-of-thought(349c2bbb-9db0-4184-bbce-9b489935520c).html
[19] Wikipedia, Multiply–accumulate operation. Available at: https://en.wikipedia.org/wiki/Multiply-accumulate_operation
[20] Fujitsu, FUJITSU Supercomputer PRIMEHPC FX1000: An HPC System Opening Up an AI and Exascale Era, Available at: https://www.fujitsu.com/downloads/SUPER/primehpc-fx1000-hard-en.pdf
[21] Wikipedia, The Millionaire Calculator. Available at: https://en.wikipedia.org/wiki/The_Millionaire_Calculator
[22] William Rae MacDonald, John Napier, Dictionary of National Biography, Volume 40, 1885-1900.
[23] Peter M.Hopp, Slide Rules: Their History, Models and Makers, ISBN 1879335867, Astragal Press, 1999.
[24] George Boole, The Laws of Thought, 1854. Available at: https://en.wikipedia.org/wiki/George_Boole
[25] Alan Turing, On Computable Numbers, with an Application to the Entscheidungsproblem, 1936. Available at: https://en.wikipedia.org/wiki/Alan_Turing
[26] Claude Shannon, A Symbolic Analysis of Relay and Switching Circuits, 1937, 1938. Available at: https://en.wikipedia.org/wiki/Claude_Shannon
[27] Claude Shannon, A Mathematical Theory of Communication, 1948. Available at: https://en.wikipedia.org/wiki/Claude_Shannon
[28] Eric Hutton, Available: http://www.englishmechanic.com
[29] Ralf Buelow, Percy Ludgate, der unbekannte Computerpionier [Percy Ludgate, the Unknown Computer Pioneer], Heinz Nixdorf MuseumsForum, 17-Jan-2020. Available in German at: https://blog.hnf.de/percy-ludgate-der-unbekannte-computerpionier/
also and including Google translation to English, courtesy Brian Randell at:
https://scss.tcd.ie/SCSSTreasuresCatalog/miscellany/TCD-SCSS-X.20121208.002/
[30] Scientific News: No. 9 of the Scientific Proceedings of the Royal Dublin Society, English Mechanic and World of Science, No. 2302, p. 322, 7-May-1909, discovered by Ralf Buelow, image provided by Eric Hutton, 19-Dec-2019.
[31] Engineering (attribution): A Proposed Analytical Machine, English Mechanic and World of Science, No. 2319, p. 111, Vol. 90 (1909/10), 3-Sep-1909, discovered by Ralf Buelow, image provided by Eric Hutton, 19-Dec-2019.
[32] Engineering: A Proposed Analytical Machine, pp.256-257, 20-Aug-1909, discovered by Jade Ward, Univ.Leeds Library, 14-Jan-2020, OCR by David McQuillan, 3-Feb-2020.
[33] Percy Ludgate, NAI/CS/PO/TR Will of Percy Edwin Ludgate, 26-Jun-1917, National Archives of Ireland.
[34] Andries de Man, Irish Logarithms, Available: https://sites.google.com/site/calculatinghistory/home/irish-logarithms-1
and https://sites.google.com/site/calculatinghistory/home/irish-logarithms-part-2-1
[35] Wikipedia, Zech's Logarithm, Available: https://en.wikipedia.org/wiki/Zech%27s_logarithm
[36] K. Hoecken, "Die Rechenmaschinen von Pascal bis zur Gegenwart, unter besonderer Berücksichtigung der Multiplikationsmechanismen", Sitzungsberichte Berliner Math. Gesellsch.13, pp.8–29, February 1913. Also available in file:
"Hoecken-MultiplierMechanisms-german-1913.pdf" at https://scss.tcd.ie/SCSSTreasuresCatalog/miscellany/TCD-SCSS-X.20121208.002/
[37] David McQuillan, Percy Ludgate's Analytical Machine, Available: http://fano.co.uk/ludgate/
[38] David McQuillan, The Feasibility of Ludgate's Analytical Machine, Available: http://fano.co.uk/ludgate/Ludgate.html
[39] Dr. Joh. Schumacher, Ein Rechenschieber mit Teilung in gleiche Intervalle auf der Grundlage der zahlentheoretischen Indizes. Für den Unterricht konstruiert, München, 1909.
[40] Dieter von Jezierski, Detlef Zerfowski, Paul Weinmann: A.W. Faber Model 366 - System Schumacher. A Very Unusual Slide Rule, Journal of the Oughtred Society, pp.10-17, Vol.13, No.2, 2004. Available: http://osgalleries.org/journal/displayarticle.cgi?match=13.2/V13.2P10.pdf also at: https://scss.tcd.ie/SCSSTreasuresCatalog/miscellany/TCD-SCSS-X.20121208.002/
[41] Andries de Man, Irish Logarithms Animation. Available: http://ajmdeman.awardspace.info/t/irishlog.html
[42] D.Riches, An Analysis of Ludgate's Machine Leading to the Design of a Digital Logarithmic Multiplier, Dept.Electrical and Electronic Engineering, University College, Swansea, June 1973.
[43] Wikipedia, Differential (mechanical device), Available: https://en.wikipedia.org/wiki/Differential_(mechanical_device)
[44] Trinity College Dublin, IBM 360/44 console and subsystems,
Available: https://www.scss.tcd.ie/SCSSTreasuresCatalog/hardware/TCD-SCSS-T.20121208.018/TCD-SCSS-T.20121208.018.pdf

47

[45] Wikipedia, Instruction pipelining, Available: https://en.wikipedia.org/wiki/Instruction_pipelining
[46] Wikipedia, PALcode, Available: https://en.wikipedia.org/wiki/PALcode
[47] Wikipedia, Epicyclic gearing, Available: https://en.wikipedia.org/wiki/Epicyclic_gearing
[48] Wikipedia, Baudot code, Available: https://en.wikipedia.org/wiki/Baudot_code
[49] Wikipedia, Gray code, Avaialble: https://en.wikipedia.org/wiki/Gray_code
[50] Maurice Manning and Moore McDowell, Electricity Supply in Ireland, The History of the ESB, Gill and MacMillan, Goldenbridge, Dublin 8, 1984.
[51] Bruce Collier and James MacLachlan, Charles Babbage: And the Engines of Perfection, pp.128, Oxford University Press, January 1999.
[52] Noel Baker, Potential €37m impact of digital life on Skibbereen, Irish Examiner, p.7, 5-Nov-2015.
[53] Bruce Collier, Little Engines That Could've: The Calculating Machines of Charles Babbage, Ph.D. Thesis, Harvard University, 1970. Available at: http://robroy.dyndns.info/collier/
[54] Brian Coghlan, An exploration of the life of Percy Ludgate, West Cork History Festival, Skibbereen, 10-Aug-2019. Available in file: "BrianCoghlan-An-exploration-of-the-life-of-Percy-Ludgate-wAnim.pdf" at https://scss.tcd.ie/SCSSTreasuresCatalog/miscellany/TCD-SCSS-X.20121208.002/
[55] Raul Rojas, The Computer Programs of Charles Babbage, Annals of the History of Computing, IEEE, March 2021.
[56] Raul Rojas, How to Make Zuse's Z3 a Universal Computer, 14-Jan-1998.
[57] Raul Rojas, Conditional Branching is not Necessary for Universal Computation in von Newmann Computers, Journal of Universal Computer Science, pp.756-768, Vol.2, No.11, 28-Nov-1996.
[58] Oscar Ibarra, Shlomo Moran, Louis Rosier, On the Control Power of Integer Division, Journal of Theoretical Computer Science, pp.35-52, Vol.24, 1983.
[59] David Harel, On Folk Theorems, Communications of the ACM, Vol.23, No.7, July, 1980.

Percy Ludgate's Analytical Machine

Brian Coghlan, Published online, 2022. © 2022, The John Gabriel Byrne Computer Science Collection

Home

The Feasibility of Ludgate's Analytical Machine

Contents

Introduction

> I have prepared many drawings of the machine and its parts; but it is not possible in a short paper to go into any detail as to the mechanism by means of which elaborate formulae can be evaluated, as the subject is necessarily extensive and somewhat complicated; and I must, therefore, confine myself to a superficial description, touching only points of particular interest or importance.

This is a small investigation into how much can be inferred about a machine designed by Percy Ludgate and described in a 1909 paper by him called 'On a Proposed Analytical Machine'. See The Computer Journal, Volume 14, Issue 3, pp 317-326 - Ludgate's analytical machine of 1909 which has a very good introduction by B.Randell to the full paper and an extract from Nature where Professor C V Boys explains how the multiplier works. Plus there's another very interesting paper by Brian Randell I learned about later which describes a similar investigation done at Swansea From Analytical Engine to Electronic Digital Computer: The Contributions of Ludgate, Torres, and Bush .

At first sight the prospects for any reasonable reconstruction seem remote. And certainly for the more advanced features I think it is unlikely one can make a sound argument for any particular choice of mechanism. However for the basic mechanism he provides overall dimensions, timings and a cursory description which do give a basis for some well founded conjecture. It was designed to be fairly small and portable so there couldn't be the incredible mass of innovations and optimisations as in the Zuse 1 which weighed about 500kg (see Konrad Zuse and His Computers).

I shall use inches everywhere as the unit of measurement. As R.A.Sutton says in 'A Victorian World of Science' about the problem of translating 'old-fashioned' units - "The problem was finally solved by ignoring it".

The Shuttles

> In my machine each Variable is stored in a separate shuttle, the individual figures of the Variable being represented by the relative positions of protruding metal rods or 'type', which each shuttle carries. There is one of these rods for every figure of the Variable, and one to indicate the sign of the Variable. Each rod protrudes a distance of from 1 to 10 units, according to the figure or sign which it is at the time representing. The shuttles are stored in two co-axial cylindrical shuttle-boxes, which

are divided for the purpose into compartments parallel to their axis. The present design of the machine provides for the storage of 192 Variables of twenty figures each; but both the number of Variables and the number of figures in each Variable may, if desired, be greatly increased.

Like Babbage he referred to the values held as 'Variables' with a capital V to distinguish them from mathematical variables which have unrestricted range.

The obvious arrangement is 64 shuttles for the inner shuttle-box and 128 for the outer with a binary selection.

Each shuttle box would need a simple mechanism to ensure the rods were held firmly in place unless being explicitly set. Nowadays we would simply have a way of just pressing the sides of the boxes to release the rods, I guess they could easily make the boxes springy enough for this.

> The machine, as at present designed, would be about 26 inches long, 24 inches broad, and 20 inches high; and it would therefore be of a portable size.

Assuming that the shuttle box takes up most of the height say 18" then a shuttle would be at most 3" wide. This means each rod must be at most about 3/21" wide. I doubt they rubbed against each other so I would guess they were either 1/8" or 1/10" wide. They would be unlikely to be much smaller than this as they would need grooves cut across them to ensure a definite digit selection at all times.

The same dimensions lead to there being 64 shuttle boxes arranged round 3x2xpi inches which gives about 18/64 or approx 1/4" for the height of the slot in which the shuttles go. Saying the metal for the shuttles and the shuttle boxes is 1/32" thick each shuttle probably took about 2x1/32" for the shuttle + 1/10" for the rods + 1/32" for the box or about 2/10" leaving about 1/4"-2/10" = 1/20" space round the shuttle. Using 1/8" wide rods would leave only 1/32" space round the shuttles but possibly the metal of the shuttles could be thinner.

The shuttle-boxes have to be at one side rather than in the middle of everything:

> It may be observed, too, that the shuttles are quite independent of the machine, so that new shuttles, representing new Variables, can be introduced at any time.

The major problem with turning the shuttles round to a particular position quickly and accurately is the play in any gearing and the torque of the shuttle boxes. He says that selecting an operand from a shuttle and adding it to the mill would take 3 seconds so the turning and stopping probably has to occupy 2 seconds or less. It is possible to do this fairly directly but I think splitting the selection into two logical parts yields a more robust solution. The first selection is of 3 or 4 bits and the second is relative to the first - both selections can be done at the same time.

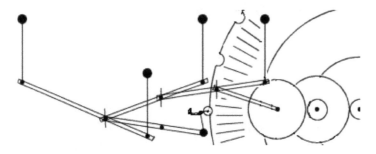

Converting Jacquard binary to shuttle-box position
(Only 4 bits of selection are shown).

Above is a simple mechanism I've devised to show how a shuttle box can be quickly and accurately positioned. The small round dots where the rods join are all constrained to move down grooves and the rods across have slots in. The conversion can also be done using differential gears but that's an unnecessary complication. I've put in a refinement to make it more robust by taking the first couple of bits move a roller on a spring backwards on the outside of the shuttle box. When the box should be in about the right position the roller is released to fit into a dip in the outside of the shuttle box thus getting the last part of the turn more accurately.

He also quite clearly had a way of calculating a shuttle box to use at run time but I've no idea how he would have done it:

> As the shuttles of the inner and outer co-axial shuttle-boxes are numbered consecutively, we may suppose the given values of a and c to be inscribed in the first and second shuttles respectively of the inner box, and of b and d in the first and second shuttles respectively of the outer box; but it is important to remember that it is a function of the formula-paper to select the shuttles to receive the Variables, as well as the shuttles to be operated on, so that (except under certain special circumstances, which arise only in more complicated formulae) any given formula-paper always selects the same shuttles in the same sequence and manner,

The options as far as I can se are:

- Have the output of the mill able to determine a shuttle directly. Besides the problems this causes in saving the contents of the mill to make up the shuttle number, this also has the problem of how does one convert the decimal mill output to a binary shuttle number.
- There could be an index register per shuttle-box holding a shuttle number. This could be set from the current box number and have binary increment and decrement operations. This would not allow one the full generality in selecting a value from a small table in a shuttle-box easily.

A too simple facility like a mechanism just to move to the next shuttle would I think have caused all sorts of trouble with using the shuttle-box for intermediate calculations. However he may have opted for a combination of the above options or something more complicated.

If the shuttle is selected using decimal rather than binary it would probably be easier to have a decimal to position mechanism rather than converting the decimal to binary first. One of Zuse's major innovations which would have simplified all this would of course have been the use of binary in the main mill.

The Index

> The fundamental action of the machine may be said to be the multiplying together of the numbers contained in any two shuttles, and the inscribing of the product in one or two shuttles.

The index is the *pièce de résistance* of the paper and it would be nice to have a good working design based on the description.

I had originally thought the shuttles would hold a straightforward representation of the digits of the Variables. However I have found if instead they hold the ordinal it cuts out a movement in the multiplier.

> Column 1 of **Table 1** contains zero and the nine digits, and column 2 of the same Table the corresponding *simple index numbers*. Column 1 of **Table 2** sets forth all

partial products (a term applied to the product of any two units), while column 2 contains the corresponding *compound index numbers*. The relation between the index numbers is such that the sum of the simple index numbers of any two units is equal to the compound index number of their product. **Table 3** is really a re-arrangement of Table 2, the numbers 0 to 66 (representing 67 divisions on the index) being placed in column 1, and in column 2, opposite to each number in column 1 which is a compound index number, is placed the corresponding simple product.

Table 1

UNIT	SIMPLE INDEX NO.	ORDINAL
0	50	9
1	0	0
2	1	1
3	7	4
4	2	2
5	23	7
6	8	5
7	33	8
8	3	3
9	14	6

The ordinals in Table 1 are not mathematically important, but refer to special mechanism which cannot be described in this paper, and are included in the tables merely to render them complete.

That the shuttles hold the ordinals rather than the units requires some justification.

- Multiplication by unity is used to do an add - using the index to convert from ordinals to units makes this more easily understood.
- The paper talks about the slides moving a simple index amount before hitting the type in the shuttle. If the units are used this would be much more difficult or would require a separate conversion stage.

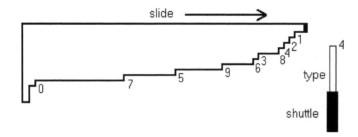

Slide hitting a type bar in a shuttle

When two Variables are to be multiplied together, the corresponding shuttles are brought to a certain system of slides called the *index*, by means of which the machine computes the product. It is impossible precisely to describe the mechanism of the index without drawings; but it may be compared to a slide-rule on which the 'usual markings are replaced by moveable blades. The index is arranged so as to give several readings simultaneously. The numerical values of the readings are indicated by periodic displacements of the blades mentioned, the duration of which displacements are recorded in units measured by the driving shaft on a train of wheels called the *mill*, which performs the carrying of tens, and indicates the final product. The product can be transferred from thence to any shuttle, or to two shuttles simultaneously, provided that they do not belong to the same shuttle-box.

Assuming 1/10" movement of the slide per unit the slides would be 5" long at least and a pair hitting each other take up 20" since they can each also move 5". If they were side by side they would still take up at least 15". The height would be at least 4" and the width 3" plus another 3" around the right hand shuttle which is shifted along. Plus of course all the space needed for the cams and springs.

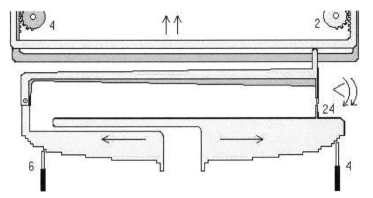

Index multiplying 4x6

The picture above omits all the little bars above the right hand slide except those corresponding to compound index number 10 which gives 24 as unit 2 and tens 4. The height of the bars is sensed by two pins each attached by a long bar to an axle at the left hand slide.

To get the next partial product cams would lift the bars and move the right hand slide inwards. The right hand shuttle would then be moved to multiply by ten (i.e. perpendicular to the picture). Springs would then pull the slide out so it hit the type. The bars on top would then be released and pulled by springs to feel the bars for the new compound index numbers.

Ludgate explains the whole business as follows:

> Now, to take a very simple example, suppose the machine is supplied with a formula-paper designed to cause it to evaluate x for given values of a, b, c, and d, in the equation $ab + cd = x$ and suppose we wish to find the value of x in the particular case where $a = 9247$, $b = 8132$, $c = 21893$, and $d = 823$.
>
> The four given numbers are first transferred to the machine by the key-board hereafter mentioned; and the formula-paper causes them to be inscribed in four shuttles. As the shuttles of the inner and outer co-axial shuttle-boxes are numbered consecutively, we may suppose the given values of a and c to be inscribed in the first and second shuttles respectively of the inner box, and of b and d in the first and second shuttles respectively of the outer box; but it is important to remember that it is a function of the formula-paper to select the shuttles to receive the Variables, as well as the shuttles to be operated on, so that (except under certain special circumstances, which arise only in more complicated formulae) any given formula-paper always selects the same

shuttles in the same sequence and manner, whatever be the values of the Variables. The magnitude of a Variable only effects the type carried by its shuttle, and in no way influences the movements of the shuttle as a whole.

The machine, guided by the formula-paper, now causes the shuttle-boxes to rotate until the first shuttles of both inner and outer boxes come opposite to a shuttle-race. The two shuttles are then drawn along the race to a position near the index; and certain slides are released, which move forward until stopped by striking the type carried by the outer shuttle. The slides in question will then have moved distances corresponding to the simple index numbers of the corresponding digits of the Variables b. In the particular case under consideration, the first four slides will therefore move 3, 0, 7, and 1 units respectively, the remainder of the slides indicating zero by moving 50 units (see Table 1). Another slide moves *in the opposite direction* until stopped by the first type of the inner shuttle, making a movement proportional to the simple index number of the first digit of the multiplier a—in this case 14. As the index is attached to the last-mentioned slide, and partakes of its motion, the *relative* displacements of the index and each of the four slides are respectively 3 + 14, 0 + 14, 7 + 14, and 1 + 14 units (that is, 17, 14, 21, and 15 units), so that pointers attached to the four slides, which normally point to zero on the index, will now point respectively to the 17th, 14th, 21st and 15th divisions of the index. Consulting Table 3, we find that these divisions correspond to the partial products 72, 9, 27, and 18. In the index the partial products are expressed mechanically by movable blades placed at the intervals shown in column 2 of the third table. Now, the duration of the first movement of any blade is as the unit figure of the partial product which it represents, so that the movements of the blades concerned in the present case will be as the numbers 2, 9, 7, and 8, which movements are conveyed by the pointers to the mill, causing it to register the number 2978. A carriage near the index now moves one step to effect multiplication by 10, and then the blades partake of a second movement, this time transferring the tens' figures of the partial products (i.e. 7, 0, 2, and 1) to the mill, which completes the addition of the units' and tens' figures thus:

$$
\begin{array}{r}
2978 \\
7021 \\
\hline
73188
\end{array}
$$

the result being the product of the multiplicand b by the first digit of the multiplier a. After this the index makes a rapid reciprocating movement, bringing its slide into contact with the second type of the inner shuttle (which represents the figure 2 in the quantity a), and the process just described is repeated for this and the subsequent figures of the multiplier a until the whole product ab is found. The shuttles are afterwards replaced in the shuttle-boxes, the latter being then rotated until the second shuttles of both boxes are opposite to the shuttle-race. These shuttles are brought to the index, as in the former case, and the product of their Variables (21893 x 823) is obtained, which, being added to the previous product (that product having been purposely retained in the mill), gives the required value of x. It may be mentioned that the position of the decimal point in a product is determined by special mechanism which is independent of both mill and index. Most of the movements mentioned above, as well as many others, are derived from a set of cams placed on a common shaft parallel to the driving-shaft; and all movements so derived are under the control of the formula-paper.

The various positions on the index slides are give by table 3 of Ludgate's paper.

Table 3

COMP INDEX NO.	PARTIAL PRODUCT	COMP INDEX NO.	PARTIAL PRODUCT
0	1	34	14
1	2	35	28
2	4	36	56
3	8	37	45
4	16	38	—
5	32	39	—
6	64	40	21
7	3	41	42
8	6	42	—
9	12	43	—
10	24	44	—
11	48	45	—
12	—	46	25
13	—	47	63
14	9	48	—
15	18	49	—
16	36	50	0
17	72	51	0
18	—	52	0
19	—	53	0
20	—	54	—
21	27	55	—
22	54	56	35
23	5	57	0
24	10	58	0
25	20	59	—
26	40	60	—
27	—	61	—
28	81	62	—
29	—	63	—
30	15	64	0
31	30	65	—
32	—	66	49
33	7		

The Mill

The sum of two products is obtained by retaining the first product in the mill until the second product is found—the mill will then indicate their sum. By reversing the direction of rotation of the mill before the second product is obtained, the difference of the products results. Consequently, by making the multiplier unity in each case, simple addition and subtraction may be performed.

In designing a calculating machine it is a matter of peculiar difficulty and of great importance to provide for the expeditious carrying of tens. In most machines the carryings are performed in rapid succession; but Babbage invented an apparatus (of which I have been unable to ascertain the details) by means of which the machine could 'foresee' the carryings and act on the foresight. After several years' work on the problem, I have devised a method in which the carrying is practically in complete mechanical independence of the adding process, so that the two movements proceed simultaneously. By my method the sum of m numbers of n figures would take $9m + n$ units of time. In finding the product of two numbers of twenty figures each, forty additions are required (the units' and tens' figures of the partial products being added separately). Substituting the values 40 and 20 for m and n, we get $9 \times 40 + 20 = 380$, or 9½ time-units for each addition—the time-unit being the period required to move a figure-wheel through 1/10 revolution. With Variables of 20 figures each the quantity n has a constant value of 20, which is the number of units of time required by the machine to execute any carrying which has not been performed at the conclusion of an indefinite number of additions. Now, if the carryings were performed in succession, the time required could not be less than $9 + n$, or 29 units for each addition, and is, in practice, considerably greater.*

*For further notes on the problem of the carrying of tens, see C. Babbage: 'Passages from the Life of a Philosopher', p. 114, etc.

The time to sum m numbers of n digits is $9m+n$ units. I'd guess there might be a small constant as well. This indicates a the use of a ripple carry. He did not have an 'anticipating carriage' doing all the carries at once by using the 9's to set predicted carry. I seems he was worried about the time per add in a multiply rather than for a single add - and he was able to completely eliminate the carry time per add - he just had a ripple carry time at the end.

The only way I can see of doing this is to use differentials to add the carries and have a screw mechanism propagating the ripple carry. From the pictures I've seen of Babbage's Difference Engine No. 2 as reconstructed by the Science Museum it looks like it does someting similar. I guess his 'anticipating carriage' is used in the mill for the Analytical Engine. I must admit to being rather surprised at how complex the carrying mechanisms in the Difference Engine look in the drawings and can't quite see why it was all needed if it is just doing ripple carries.

The timing indicates that the next partial product was being set up whilst the last was still being added into the mill. Presumably the tens had an intermediate register which was set up whilst the units were being added in, then the intermediate register was added in whilst the next partial product was set up.

The Keyboard

The Analytical Machine is under the control of two keyboards, and in this respect differs from Babbage's Engine. The upper key-board has ten keys (numbered 0 to 9), and is a means by which numbers are communicated to the machine. It can, therefore, undertake the work of the number-paper previously mentioned. The lower key-board can be used to control the working of the machine, in which case it performs the work of a formula-paper.

The paper specifies only one set of numeric keys 0-9 instead of one per digit. I'd probably have gone for a set of slides but that's just taste. I would guess the partial products mechanism was used to place the digits one at a time into the mill.

The selection of the shuttles probably had keys marked with binary numbers as he had the idea of them being in sequence - and this would give an easy way of converting from decimal.

The action of the keyboards would cause holes to be put in the formula paper. As well as the two keyboards there would also be a printer:

> The machine prints all results, and, if required, the data, and any noteworthy values which may transpire during the calculation. It may be mentioned, too, that the machine may be caused to calculate and print, quite automatically, a table of values—such, for instance, as a table of logs, sines, squares, etc. It has also the power of recording its results by a system of perforations on a sheet of paper, so that when such a *number-paper* (as it may be called) is replaced in the machine, the latter can 'read' the numbers indicated thereon, and inscribe them in the shuttles reserved for the purpose.

I think the number paper and formula paper were simply two names for the same thing but used differently, as he states he differs from Babbage in not having separate inputs for the formula and the numbers. It is a bit confusing having the two terms though.

The Formula Paper

> Babbage's Jacquard-system and mine differ considerably; for, while Babbage designed two sets of cards—one set to govern the operations, and the other set to select the numbers to be operated on—I use one sheet or roll of perforated paper (which, in principle, exactly corresponds to a set of Jacquard- cards) to perform both these functions in the order and manner necessary to solve the formula to which the particular paper is assigned. To such a paper I apply the term formula-paper. Each row of perforations across the formula-paper directs the machine in some definite step in the process of calculation—such as, for instance, a complete multiplication, including the selection of the numbers to be multiplied together. Of course a single formula-paper can be used for an indefinite number of calculations, provided that they are all of one type or kind (i.e. algebraically identical).

The paper very definitely talks about using paper rather than cards as in a Jacquard Loom. Therefore the machine would have had to have a metal platen with holes to support the paper. The platen could also be used to lift up the rods that passed through as in a Jacquard loom, this would avoid tearing the edges of the holes in the paper and allows force to be exerted on whatever the holes control.

Each row does one step, so there would have to be:

- 7 holes to select a Variable in the outer shuttle box
- 6 holes to select a Variable in the inner shuttle box
- 2 holes to say whether the shuttles are selected
- 11 holes for a digit or sign entry from the upper keyboard
- Probably one hole per command rather than an encoding
- Maybe some skip marker holes for conditional skip targets

He clearly intended to have a way of dealing with conditions but doesn't say much about it. I would guess he would have it skip to a skip marker like in the old printers rather than try to skip a certain number of instructions:

> It can also 'feel' for particular events in the progress of its work—such, for instance, as a change of sign in the value of a function, or its approach to zero or infinity; and it can make any pre-arranged change in its procedure, when any such event occurs. Babbage dwells on these and similar points, and explains their bearing on the automatic solution (by approximation) of an equation of the nth degree ;* but I have not been able to ascertain whether his way of attaining these results has or has not any resemblance to my method of so doing.

> ***C. Babbage: 'Passages from the Life of a Philosopher', p. 131.**

This add up to something like 40 holes per line I think which at 1/10" per hole and 1/10" between them makes the paper at least 8" wide, a bit too wide for me to classify as a 'tape' but he may merely mean it is a continuous roll.

The Division Table

$$(2) \qquad \frac{p}{q} = Ap(1-x)(1+x^2)(1+x^4)(1+x^8), \text{etc.}$$

The series (1) converges rapidly, and by finding the sum as far as x^{10} we obtain the correct result to at least twenty figures; whilst the expression (2) gives the result correctly to at least thirty figures. The position of the decimal point in the quotient is determined independently of these formulae. As the quantity A must be the reciprocal of one of the numbers 100 to 999, it has 900 possible values. The machine must, therefore, have the power of selecting the proper value for the quantity A, and of applying that value in accordance with the formula. For this purpose the 900 values of A are stored in a cylinder—the individual figures being indicated by holes of from one to nine units deep in its periphery. When division is to be performed, this cylinder is rotated, by a simple device, until the number A (represented on the cylinder by a row of holes), which is the reciprocal of the first three figures of the divisor, comes opposite to a set of rods.

I find the division table the least convincing part of the whole design. My major concerns are how it could be manufactured and kept clean and usable, and how one could quickly select a particular line.

The division table would consist of 900 numbers of 20 digits arranged as pits on a wheel. Supposing an 20" diameter wheel is used this gives about 60" round the circumference or 900/60 = 15 numbers per inch. Obviously the pits could not accept normal width rods. The absolute minimum size I could see being used for feeler pins would be about 1/32". My guess therefore would be that the holes would be 1/20" wide to avoid the pins sticking and also avoid the pits running into each other which could make them hard to drill.

It is possible to put more than one selection per row so one has a long cylinder instead of a wide wheel but I'm not certain that gains very much.

Manufacture of the wheel would involve drilling 20x900 = 18000 holes as one couldn't form it from a mould. Well I suppose one could use a mould if one used rubber like in a modern day toy music box. Without using a mould there would have to be some sort automatic copy from a master wheel. I wouldn't fancy making the master - it would take 18000/60/60 = 5 hours at one hole per second and one certainly wouldn't be working that fast.

Operation of the wheel would involve selecting one of the 900 positions and pushing in feeler pins to get the value out. The selection does not have to be as fast as for the shuttle boxes but the torque would be much higher because of the solid rim and the accuracy required would be much greater.

I really can't see why he couldn't be happy with 90 numbers instead. The division would only take two extra multiplies and the cylinder would become fairly manageable.

So after that rather negative account of the division table you can guess my reaction to where he goes on with:

> It will be observed that, in order to carry out the process of division, the machine is provided with a small table of numbers (the numbers *A*) which it is able to consult and apply in the proper way. I have extended this system to the logarithmic series, in order to give to that series a considerable convergency; and I have also introduced a *logarithmic cylinder* which has the power of working out the logarithmic formula, just as the dividing cylinder directs the dividing process. This system of auxiliary cylinders and tables for special formulae may be indefinitely extended.

Motor and Cams

> While the machine is in use its central shaft must be maintained at an approximately uniform rate of rotation—a small motor might be used for this purpose. It is calculated that a velocity of three revolutions per second would be safe; and such a velocity would ensure the multiplication of any two Variables of twenty figures each in about 10 seconds, and their addition or subtraction in about three seconds.†
>
> **†The times given include that required for the selection of the Variables to be operated on.**

The motor would have been separate and not included in the box. This sort of arrangement was quite common at the time - the motor could be put on the floor or under the table and connected by a driving band.

Having a near constant speed rather than being worked by hand solves a number of problems with the operation of fast return mechanisms and springs.

D. RICHES

An analysis of Ludgates machine
leading to a design of a
digital logarithmic multiplier.

department of electrical and electronic engineering

university college of swansea   wales   united kingdom

122

university college of swansea

wales   united kingdom

department of electrical and electronic engineering

author   D. RICHES

title   An analysis of Ludgates machine
leading to a design of a
digital logarithmic multiplier.

supervisor _____

approval _____

date _____

Abstract:

    This paper documents an analysis of a design by
Percy Ludgate, during the years 1903-1909, of a
program-controlled mechanical calculator, or analytical
engine. The machine is then compared with its modern
counterparts, leading to a design of an electronic
digital multiplier using a logarithmic principle.

CONTENTS

123

1

INTRODUCTION

A short account by Ludgate on his analytical machine appeared in the Scientific Proceedings, Royal Dublin Society, April 1909. A paper documenting a search for further information on Ludgate and his machine was carried out and documented by Randell in 1971. [1] No new information concerning the machine's design was discovered, making the 1909 paper the only source of information on which this report could be based.

Ludgate is rarely mentioned in the introductions to standard computing texts, with the result that they jump from Babbage's work (started 1834, ended 1871) to that of Aiken (started 1931) when considering program-controlled calculators.

One purpose of this paper is to document an analysis of Ludgate's machine after an attempt to 'expand' the information in the report.

The design is then compared to conventional systems with a view to determining any aspects of the machine not yet used in modern machines. This leads to a new design of digital electronic multiplier based on the logarithmic principle used in Ludgate's design.

Finally the practicability of this type of multiplier and further applications of the logarithms are discussed.

2

SYMBOLS

A_n)
)
) Adder inputs to nth stage
B_n)

C_n carry from nth stage adder

CSA   carry-save-adder

LS   least significant

MS   most significant

ns   nanosecond

Σn sum from nth stage adder

See Appendix C for logic symbols

CHAPTER ONE

HISTORY OF CALCULATING MACHINES

The intention of this section is to give an outline of the major advances in calculating machines, prior to the publication of Ludgate's paper in 1909. This has been summarised in a time chart (figure 1).

The examination and reading of calculating machine designs [ref. 2 to 10] greatly helped in the analysis of the proposed analytical engine.

The early digital mechanical calculators can be divided into three groups, desk calculators, difference engines and analytical engines. The desk calculators were intended to reduce the time and errors of simple calculations and were not automatic. The earlier type of this calculator, such as Pascal's, could only add or subtract. In 1671 Leibniz produced a design for a calculator that could also multiply. It was not until 1820 that the first calculator capable of all four basic arithmetic operations was made for commercial manufacture. This was the 'Arithmometer' of C.X. Thomas. The 'Millionaire', designed by Steiger, was a more notable early success and was again capable of the four arithmetic operations. The 'Millionaire' and a machine by Bollé were unusual examples of calculators in that they performed multiplication by a direct method, and not successive addition as in most other machines [4]. In these machines the operands had their separate digits multiplied by what was basically a table look-up

system.

Around 1786 a new type of calculator design was conceived by Muller. This was the difference engine. In 1812, probably unaware of Muller's work, Charles Babbage began the design of a difference engine and in 1822 completed a working model. A difference engine is only suitable for the automatic calculation of mathematical tables of functions whose higher order differences are constant. The engine or machine comprises a register for each order of difference and mechanism for adding the data in each register to that of the next lower register. A concise description of the principles of such a machine has been written by Babbage [2]. Despite its simplicity of action, it was not until around 1854 that a difference engine, with a useful number and size of registers, was constructed. This was designed and built by Scheutz. Babbage never completed the construction of a full-scale version of his difference engine and in 1833 turned his attention to designing an analytical engine.

Babbage's analytical engine was to be more powerful than any difference engine, with the results of calculations being able to affect the future instructions of the machine. It was to be capable of automatically computing any algebraic formula for which there was a solution, given the initial values. The machine was to be controlled and data communicated to it by punch cards.

The principles of punch card controlled machinery were first demonstrated by Bouchon between 1725 and 1745. By 1804 Jacquard was using this method of control for weaving cloth automatically. A short account of punch card control appears in Morrison [9].

5

There are several descriptions of the principles of Babbage's analytical engine [2, 7, 8]. The original drawings and parts of the machine can be seen at the Science Museum, London. The designs for the engine included a mill where the arithmetic operations of addition, subtraction, multiplication and division were to be performed. Numerical data was to have been stored on 1000 columns of wheels. Each number was to have a separate column and the position of rotation of each wheel was to represent a separate digit. The numbers were to be to 50 decimal places. Babbage intended to use two different sets of punched cards for conveying numerical data and algebraic formulae (program) to the machine. Shortly before his death Babbage built the arithmetic and printing units. His son, Henry Babbage, continued working on the analytical engine after his father's death but it was never completed.

It should be realised that at the time Charles Babbage was designing and building his analytical engine, desk calculators with four arithmetic operations had not yet reached any notable commercial success.

It was in 1909 that Ludgate published what appears to be his only report concerning his design of an analytical engine [11]. He mentions here that in the early stages of his work he had no knowledge of Babbage's engines and that later on he only had a slight knowledge of them, limited mainly to their mathematical principles. It is clear from reading reports on Babbage's and Ludgate's analytical engines that they are mechanically completely different and that the principles behind each can only be compared on a general level.

6

Ludgate states briefly in a later paper [7] that he had nearly completed a second design,

" .... in which are combined the best principles of
both the analytical and difference types, and from
which are excluded their more expensive characteristics."
No further information on this design can be found.

From this summary of calculating machines it can be seen that Ludgate's work appears original and worthy of investigation. The following two chapters of this report refer in detail to his first design of an analytical engine.

CHAPTER TWO

DESCRIPTION OF LUDGATE'S ANALYTICAL MACHINE

7

2.1 DEFINITION OF AN ANALYTICAL ENGINE

It is convenient to start with a more detailed description of what Ludgate considered the necessary operations and facilities of an analytical engine. This will give a clearer picture of what Ludgate was designing.

In his 1909 paper Ludgate wrote that the object of his work was to design

"machinery capable of performing calculations, however intricate or laborious, without the immediate guidance of human intellect."

The required operations of an analytical engine can be extracted from his report and can be enumerated as follows:

1. A form of communication between machine and operator.

2. ".... means of storing the numerical data of the problem", plus the intermediate results and final answer(s).

3. Capacity to submit ".... any two of the numbers stored to the arithmetical operation of addition, subtraction, multiplication or division."

4. The ability ".... to follow a particular law of development as expressed by an algebraic formula."

5. The ".... changing from one formula to another as desired, or in accordance with a given mathematical law."

6. The capacity to ".... 'feel' for particular events in the progress of its work".....'" and also to make any pre-arranged change in its procedure, when any such event occurs."

8

It is worthwhile noting at this stage that these basic requirements also occur, but in a different form, in a much later report (1946) by von Neumann, Goldstine and Burks concerning "Preliminary discussion of the logical design of an electronic computing instrument." [12]

2.2 MATHEMATICAL THEORY BEHIND LUDGATE'S MACHINE

The fundamental action of Babbage's machines was addition, whereas in Ludgate's it was to be "direct" or "partial product" multiplication. In a machine whose fundamental action is addition, subtraction is performed by reversing the process, multiplication by successive addition and division by successive subtraction. In contrast the basic action of the arithmetic unit in Ludgate's machine was to be multiplication by a logarithmic method.

In his report he mentions that he originally intended to use ordinary logarithms to base ten,

"... but found that some of the resulting intervals were too large, while the fact that a logarithm of zero does not exist is an additional disadvantage."

9

He therefore arranged for each of the prime numbers below ten to have associated with it an index number. The indexes or logarithms of non-prime numbers, i.e. all possible products of prime numbers up to 9 x 9 = 81, are formed by adding the index of the prime numbers that form that product. Appendix 'A' contains an algorithm for writing such a system of logarithms.

The index numbers of the base ten system that Ludgate used are shown in table 1. The use of these index numbers in multiplication can be explained as follows. When two single digit numbers are to be multiplied then the corresponding simple index numbers are added. This result is called a compound index number and by referring to table 2 the corresponding product is found. Two examples are shown in table 1.

Multiplication of numbers of more than one digit is broken down into multiplication of single digits. An example of this, using tables 1 and 2, is shown in table 3. The allowance for the factors of ten in the multiplier and the 'carry' digits in the partial product addition were to be mechanical.

Addition was to be performed by using the same mechanism as in multiplication and was to be effected by multiplying the addendums by unity. These products were then to be summed as were the partial products of a multiplication. Subtraction was similar to addition except in the final stage. Here a train of gears for accumulating the partial products was to be rotated in the reverse direction.

Division was to be performed in what was then an equally unusual way as the multiplication [13]. The basis of the scheme was to

10

multiply the dividend by the reciprocal of the divisor. The reciprocal was first to be estimated by what may be considered a table look-up system. This estimate was then to form the basis of a highly convergent series in which only addition, subtraction and multiplication would be required to solve its sum.

The theory of the division is as follows:

Assume the Quotient $= \dfrac{D}{d}$

The reciprocal of the three most significant digits of 'd' is found by table look-up. Let this be 'A' (In the proposed machine A was to be a decimal of 20 figures.)

$\Rightarrow$ A.d begins with the decimal digits 100 ...

Let A.d be of the form 1.00 ... or $1 + x$ where 'x' is the small fraction

Then $A.d = 1 + x$

$\therefore \dfrac{1}{d} = \dfrac{A}{1+x} = A(1+x)^{-1}$

Expanding by the binomial theorem

$\dfrac{1}{d} = A(1 - x + x^2 - x^3 + x^4 - \ldots \pm x^n \mp \ldots)$

$= A(1 - x)(1 + x^2)(1 + x^4)(1 + x^8) \ldots (1 + x^{2n-2}) \ldots$

Therefore

Quotient $= \dfrac{D}{d} = D.A(1 - x)(1 + x^2)(1 + x^4) \ldots$

By taking this last series as far as x^{10} the result is given correct to at least thirty figures. Ludgate states that the position of the decimal point was to be found independently of the formula and by a mechanical method. This was to be so for multiplication, addition and subtraction also.

The numerical data was to be represented in the form of twenty digit variables with an extra sign digit (i.e. sign magnitude representation).

2.3 MECHANICAL DESCRIPTION

Randell's search for information [3] concerning the analytical engine suggests that Ludgate never succeeded in having his machine constructed. The drawings and manuscripts have not been traced despite a thorough search. Thus the 1909 paper is at present the only source of information on the machine and this was only written to serve as a short account of his work. From his report it has been possible to derive sketches of the machine, though there is no way of telling at present how closely these drawings follow the original design. They do, however, correspond exactly to the description in Ludgate's report and also serve in the understanding of the machine.

The Store

Ludgate intended his machine to have a store of 192 variables of twenty digits plus a sign digit. Each variable was to be stored in a shuttle, the individual digits being represented by protruding rods (see figure 2(a)). Two "co-axial cylindrical shuttle-boxes," divided into compartments parallel to their axis, were to hold the 192 variables. The inner and outer shuttle-boxes would each contain 96 shuttles, (see figures 2(b) and 2(c)) and the store would therefore be divided into two distinct sections.

Ludgate states that both the number of variables and the number of digits in each could be increased. Also he mentions that new shuttles, representing new variables, could have been introduced after removing the old ones. To access a variable the store was to be rotated until the variable was brought opposite a 'shuttle-race' or track. The shuttle was then to be drawn out along the race. There were two 'shuttle-races' in the design, one for the outer store and one directly below for the inner store. Thus two variables on the same axis but different stores could be accessed simultaneously. This can be seen more clearly in figure 2(c).

It appears from considering aspects of the analysis not yet mentioned, that when programming the machine, any two variables to be multiplied may have to have been stored on the same axis. There is no suggestion in the report that two variables on different axes in different stores could be multiplied directly. This may have been possible, however, by taking a shuttle from one of the stores, rotating the stores, and taking another from the other store. The variables would have been multiplied and the shuttles returned to their respective locations. This type of operation could have been avoided but storage would have been wasted. The machine's capacity to do this would have determined its control mechanism complexity.

The Arithmetic Unit

A system of slides, called an index, was to convert the variable on the rods of an outer shuttle into distances corresponding to the simple index numbers of that variable. These slides were probably

11

12

D. Riches, Department of Electrical and Electronic Engineering, University College, Swansea, U.K., 1973. © 1973, University College Swansea, reproduced with permission

An analysis of Ludgates machine leading to a design of a digital logarithmic multiplier

13

to be stepped as shown in figure 3(c) and were to be in line before operation as in figure 3(a). They were to be released and moved forward until stopped by striking the 'type' of a shuttle as in figure 3(b). The distances moved are logarithmic displacements of the twenty digits in the shuttle.

In the design, the protrusion of the rods in a shuttle is governed by table 4. The slides are stepped in distances corresponding to the simple index numbers, as shown in figure 3(c).

Figure 4 is an overall sketch of the arithmetic unit of the machine and should be referred to when reading the following description.

The design includes only one bank of slides, running against the rods of an outer shuttle. A single slide moves in the opposite direction to these slides and runs against the most significant digit of an inner shuttle. This single slide is attached to the index, which then also moves a distance corresponding to the simple index of the most significant digit of the multiplier. The resultant displacements of the slides are compound index numbers, (i.e. simple index number of most significant digit of the inner variable added to each of the index numbers of the digits of the outer variable).

It is not clear in the report how Ludgate intended to convert these relative displacements into the partial product. He states:

"The numerical values of the readings" (compound index numbers)"are indicated by periodic displacements of the blades mentioned, the duration of which

14

displacements are recorded in units measured by the driving shaft on a train of wheels called the mill".

After these compound index numbers had been converted to the partial product and added to the mill mentioned above, the process was to be repeated again. The single index rod then runs against the second most significant digit of the multiplier. The mill, being a train of gears, was to be capable of allowing for the carrying of tens when summing the partial products and allowing for multiplication by ten each time a partial product is added.

There is no mention of how the results in the mill were to be returned to the store. Figure 4 includes a possible method, again using stepped bars.

Multiplication of two twenty digit numbers was to have taken ten seconds and division about 1½ minutes maximum. Add and subtract were multiplication by unity. It appears from timings given in the report, that to add or subtract necessitated the variable being in the outer store. These operations were then to have taken a time of three seconds each. If it was possible for the variable to be in the inner store, it would have been multiplication of unity by the variable with a time corresponding to that of an ordinary multiplication.

There is no mention of how the machine was to deal with the double length words that may be produced in the arithmetic unit. Presumably there was to be no provision for numbers greater than twenty decimal digits. Neither does the report indicate whether overflow of this type could be indicated by the machine and how such a number was to be reduced to a single word length.

Control, Input/Output

Ludgate's method of control was similar to Babbage's in that they both used a Jacquard system. Babbage was to use perforated cards while Ludgate designed his machine to use perforated tape or 'formula-paper'. The information in the perforations or holes was to be converted to mechanical displacements by rods. The absence or presence of holes was to determine the instruction in mechanical movements. One function of this 'formula-paper' was to select where the variables were to be stored and which shuttles were to be operated upon. Each row of perforations was to direct the machine one step of a calculation, i.e. a complete multiplication including the accessing of the two variables.

For calculations that did not warrant a 'formula-paper', the design included a keyboard. This was to control the machine and also to act as a means of punching a new 'formula-paper'.

A second keyboard was included by which numbers were to be communicated to the machine. The machine was also to be able to produce a 'number-paper' which could record numerical data. This could then be replaced in the machine and the data re-entered.

To avoid having to repeat a series of instructions on the 'formula-paper' every time a divide instruction occurred, Ludgate devised a subroutine. When a division was indicated on the 'formula-paper', control was to be passed to a dividing-cylinder. This cylinder was to contain the required perforations.

The reciprocal of the three most significant digits of any divisor was to be found from a cylinder, the 900 possible values being represented as twenty digit numbers in the form of holes one to nine digits deep. Rods were to transfer the reciprocal to a shuttle.

A logarithmic cylinder was also to be included. This in principle was similar to the dividing cylinder but with the perforations for calculating a logarithmic formula. Ludgate intended his machine to accept control cylinders of different functions.

One of the powers of the machine was to be its ability to change control from one formula to another "in accordance with a given mathematical law." It was to "feel" for events in calculations such as changes in sign or approaches to infinity. Ludgate only mentions this briefly. This form of conditional branching was probably to have been accomplished by skipping specified rows of instructions on the 'formula-paper'.

There is no way of determining exactly the instructions that were to control the machine. This is largely due to not being able to determine which pairs or single shuttles can supply variables for an arithmetic operation.

The ingenuity of the design can be appreciated when the size of the proposed machine is considered. Ludgate gave the dimensions as 26 inches long, 24 inches broad and 20 inches high. This is considerably smaller than the engine designed by Babbage which would have been measured in feet.

17

CHAPTER THREE

COMPARISON OF LUDGATE'S MACHINE WITH ITS
MODERN COUNTERPARTS

One purpose of comparing Ludgate's machine with its modern
counterparts is to determine whether there is any undeveloped
aspect of his design which could be used in a modern computer.

3.1 LUDGATE'S MACHINE TODAY

Firstly the information about the machine has to be put into a
form that can be compared with modern designs. The block diagram
(figure 5) gives a basic layout of the engine. The control sections
are omitted since they cannot be precisely defined. Each 'block'
in the diagram is under the direction of a control system which in
turn receives instructions from the keyboard or 'formula-tape'.
It is not possible to judge whether the input/output of numerical
data was to be via the accumulator or direct to the store.

The store is represented as two separate units. The removal
of the shuttles from the store to access the data has been represented
by a memory buffer. The register I_0 corresponds to the slides of
the index which were to effectively store the variable by their
displacements, while multiplication took place.

An example of what one line of perforations across the
formula-paper could direct the machine to do, is the selection of

18

two variables and the multiplication of them. This is stated in the
report. A set of instructions on which the engine could operate
has been derived and tabulated in table 5. It is possible to have
such a machine working on the operation codes one to eleven in the
table. This restricts its multiplication/division to variables
with the same address but in different stores, (i.e. same axis).
This may have been overcome by data being transferred around the
store via the mill, though this process would have been very slow.

Operation codes twelve to fifteen allow temporary storage of
data on the index slides and a temporary storage register 'T' common
to both stores. These would have permitted faster arithmetic
operations on variables stored anywhere.

The format of a possible instruction word is given in figure 6.
This shows the store as being two interleaved stacks with double
addressing for adjacent locations, i.e. two bit store address and
seven bit word location address.

Figure 7 gives a flow chart of the sequence of stages in such
a machine for the multiplication of two variables with the same
address but different stores. An explanation of this is given
opposite figure 7. Figure 9(a) gives the instruction words for
such an operation following path ❋ of the flow chart. Multiplication
of variables stored in different formations is explained in figure
8. From figures 7 and 8, it is obvious that for the lowest
multiplication time, the variables must be stored with the same
address in each store. This enables them to be accessed simultaneously.

D. Riches, Department of Electrical and Electronic Engineering, University College, Swansea, U.K., 1973. © 1973, University College Swansea, reproduced with permission

An analysis of Ludgates machine leading to a design of a digital logarithmic multiplier

3.2 COMPARISON

In the conclusion to his report and by considering the previous chapters it is clear that Ludgate intended his machine for mathematical use. Therefore it should be considered an 'early scientific computer'.

The basic layout of the machine closely follows that of today's machines as did Babbage's design. They all have arithmetic, control, input, output and store units.

Store

Besides the main store, the machine was to have a read only store for finding the reciprocals in the division routine. This can be described as a predecessor to some of today's read only memories.

It appears that Ludgate was attempting to reduce the number of separate store accesses by dividing the main store in two and and enabling double access. This facility has been included in the Atlas machine [14]. Here the core is split into four stacks, each having its own read, write and decoding mechanisms with a page of instructions spread over two stacks. Due to this, it is possible to read a pair of consecutively stored instructions in parallel. Another machine that Ludgate's engine can be compared to in this aspect is the I.B.M. 7094. This has a facility for requesting two 36-bit words as the actual memory has a 72 + 1 bit parity word for even and odd addresses [15]. The Control Data 'Star' computer has the facility of removing several operands from its store simultaneously.

The total capacity of 192-variables seems very low, although it was mentioned that this could have been increased. There was provision in the design for perforated paper storage outside the machine which can be compared to magnetic tape storage of today.

The proposed data word length of twenty decimal digits (requiri sixty bits for binary representation) is large even by today's standards. It is comparable with the I.B.M. Stretch computer [16] a the early electromechanical machine, Harvard Mk 1 which had twenty-four wheels representing twenty-three decimal digits and a sign digit.

The cyclic access principles of Ludgate's store were used in most vacuum tube computers but today they form the basis for some types of backing store (e.g. disc).

The removing of shuttles from the store to the index may be considered a destructive read. The variable must be re-written into store (i.e. shuttle replaced).

The advantages for Ludgate of storing instructions in the main store, as in most electronic computers, would have been greatly outweighed by the mechanical disadvantages of complexity and relativ slowness.

The Arithmetic Unit

Multiplication in modern computers is normally performed by repeated addition. This is similar to the 'pencil and paper' method The basic operations of these machines are addition and subtraction. Ludgate's design for a logarithmic multiplier appears to be unique. The potential of this principle is examined, with regard to modern

21

technology, in the following chapters.

Subroutine methods of division by finding the reciprocal of the divisor have been discussed in several papers [17, 18, 19, 20, 26] . In 1946 von Neumann considered the use of an iterative scheme by this method. This was compared to a hardware design with the resulting recommendation that the hardware divider was built. This was due to cost/speed considerations. For Ludgate to have included a separate mechanical divider in his machine would have added to the cost and complexity considerably.

Flynn [17] describes an identical routine to that proposed by Ludgate for determining the reciprocal, while Wallace [18] and Ferrari [19] describe a more common approach using a Newton-Raphson iteration. Both methods require an initial approximation to the reciprocal. The accuracy of this approximation determines the number of stages of calculation needed to find the result to the required accuracy. Ludgate's method of table look-up to determine the initial approximation has been used in the I.B.M. System/350 Model 91 [21] . The division routine in this I.B.M. machine is discussed, and improvements suggested, by Ahmad. [26] The routine is identical to Ludgate's in theory, although it is applied to bit normalised binary number system.

Control, Input/Output facilities

In Ludgate's machine, these bear a strong resemblance to modern methods although they would have been primitive in most aspects.

His design included two paper-tape readers; formula-tape and number-tape. Thus instructions and data were kept completely separate.

22

These readers were also designed to produce new punch tape, thereby creating a backing store. In modern machines, technology has enabled these two types of punch to be combined into one unit.

It is not possible from the report to determine which unit (store or mill) was to accept or output data. Thus, no comparison can be made here although both methods have been used since.

His design permitted keyboard control of the machine and of a printer. This is a common facility today. The method mentioned earlier of conditional branching is primitive by today's standards, though it does compare with the conditional branching in which is now considered the first computer, the Harvard Mk 1 (1944).

3.3 CONCLUSION

Judging from the limited evidence provided by Ludgate's 1909 report, it would appear that in theory many features of his design are similar to equivalent aspects of modern computers. On the other hand there is little mention of practical considerations, and Ludgate himself states in his later report [7] that

"the true calculating machine belongs to a possible rather than actual class."

An analysis of Ludgates machine leading to a design of a digital logarithmic multiplier

CHAPTER FOUR

ELECTRONIC LOGARITHMIC MULTIPLIERS

As noted earlier one unusual aspect of Ludgate's machine is his use of logarithms to effect multiplication. In this section it is intended to describe the development and design of an electronic multiplier based on the same principle as that proposed by Ludgate. The design was considered for construction in Transistor Transistor Logic (T.T.L.)

4.1 CHOICE OF LOGARITHMS

Ludgate wrote his logarithms to base ten, (decimal). This type of logarithm may be written to any base, (see Appendix 'A'). It is important when calculating these logarithms to make the simple index numbers as low as possible. This reduces the complexity and propagation delay time of an electronic multiplier when used as a basis for its design.

By writing the logarithms to base eight or base four it becomes possible to convert a binary number directly into its logarithm. This is done by examining and converting to indexes in three or two bit groups respectively.

With the base eight system a four digit octal word (12-bit binary)

can have its digits converted to simple index numbers in four separate parallel stages. This makes it possible to multiply two twelve bit numbers in four cycles by multiplying by three bit groups. A similar approach but using base four logarithms will require six cycles.

4.2 A MULTIPLIER USING LOGARITHMS TO BASE EIGHT

The logarithms or index numbers written to base eight are contained in tables 6 and 7. A general outline of a multiplier using these logarithms will be considered first.

General Description

A schematic plan of the multiplier is given in figure 10 and a labelled path through part of it in figure 11. The design is based on the multiplication of two twelve bit binary numbers. The design can easily be modified for other word lengths. It is more suited to numbers with bit lengths that are multiples of three as it 'views' three bits of the multiplier in one cycle.

The action of the multiplier can be described in the following sequences:

1. The four octal digits of the multiplicand and the least significant octal digit of the multiplier are converted in five parallel stages to their simple index numbers.

2. The simple index number of the multiplier is added to each of those formed from the multiplicand. Thus four compound index numbers of six bits each are produced.

25

3. These six bit compound index numbers are converted to six-bit (two digit octal) semi-partial products.

4. The most significant octal digit (3 bits) of the semi-partial products are added to the least significant digit of the next highest semi-partial product. This is similar to the procedure described by Ludgate for his machine.

5. The result of the previous stage may be considered a complete partial product since it is the product of multiplicand and three bits of the multiplier. This partial product is stored in a parallel load shift register.

6. The complete partial product and multiplier are now shifted right three bits. This shift aligns the next significant three bits of the multiplier ready for the next cycle. The shift of the complete partial product is to compensate for the multiplication factor of eight. (This is comparable to the multiplication by ten in the mill of Ludgate's machine).

7. The cycle is now repeated with the next three bits of the multiplier. The complete partial product formed by this cycle is added to the previous one and the sum and multiplier simultaneously shifted right three bits. Four cycles are required giving a final product of twenty four bits.

26

The connections through a path of the multiplier as shown in figure 11 have been labelled to correspond to the following descriptions.

Storage Register for Multiplier and Multiplicand

The multiplier and multiplicand are held in twelve bit parallel load/out registers, (assuming parallel mode of machine structure). Twelve binary latches are required for the multiplicand and a shift right register for the multiplier. It is advantageous in the next stage that both these registers should be able to supply the complement of the number stored.

Octal Digit to Simple Index Converter

This logic circuit converts a three bit unit number into its simple index number which has a maximum range of six bits. (The six bits are required to accommodate the index of zero). One important consideration in the design is to produce a circuit with a minimum delay. Figure 12 is such a circuit with a propagation delay of two gates.

Simple Index Adders

Four six bit adders are required to add the simple index numbers. Since the length of the index numbers is only six bits, medium scale integrated (M.S.I.) circuit adders with ripple carry may be used. (Carry anticipate circuits compare with short M.S.I. adders in speed, [?3], this can be verified for a six bit adder using propagation delay data of 'Texas Instruments' components. [22]. The disadvantage of a more complicated circuit outweighs the advantage of an approximate five nanosecond reduction in delay.)

A seven bit number from the adders can only be produced if the index of zero is added to itself, (i.e. 0 times 0.) This gives a compound index result of octal 112. It is fortunate that in the logarithms to base eight there is no way in which a compound index of octal 12 can be produced. Therefore by ignoring the most significant bit of octal 112 and associating semi-partial product zero with compound index number 12, the semi-partial product is reduced to six bits.

Compound Index Number to Semi-Partial Product, Converter

This section of the logic converts the six bit compound index number (K, L, M, N, P, Q, see Fig. 11) to a six bit semi-partial product. (R, S, T, U, V, W) Again a main consideration is to produce a circuit with a minimum propagation delay.

A compromise between circuits with varying numbers of eight, four and three input logic gates has to be made.

The most significant bit of the input can be ignored for the first part of the design. A binary 'one' only occurs here for output zero and two other values. This reduces the Karnaugh Map analysis to five input variables. The Karnaugh Maps for this conversion are included to show the 'awkwardness' of the system (Table 8). The Boolean equations for the outputs, assuming a logic zero for the sixth input, are also included. This most significant sixth input is included in the final logic diagram, figure 13. The circuit takes into account the fan-out (max. 10, [22]) of the logic gates. This design requires 64 gates of which ten are eight input and has a maximum propagation delay time of four gates.

27

Without any optimisation in the design, a suitable logic circuit requires 69 gates, 31 of them being of the eight input type. It also has a maximum delay of five gates.

Four of these converters are required.

Semi-Partial Product to Complete Partial Product
(and addition of Complete Partial Products)

The operation of this section of the circuit can be described in five steps, in conjunction with figure 10.

1. The three most significant bits of the previous stage outputs added to adjacent least significant bits.

    i.e. outputs from previous logic

    m.s. - RSTUVW - l.s.

    added in following manner

    $$R_n + U_{n+1}, \quad S_n + V_{n+1}, \quad T_n + W_{n+1}$$
    $$\text{where } n = n\text{th stage}$$

2. Above result added to previous complete partial product. (Ripple carry adder bank.)

3. Result 'accepted' into parallel in/out shift register of twenty four bits.

4. Result and Multiplier shifted right three bits.

5. Copy into twelve bi-stable latches, which have outputs connected to second bank of adders, enabling sum of partial products to be added on next cycle.

There are several possible designs for this part of the circuit.

28

29

Figure 10 contains the simplest which is two banks of ripple carry
adders. The carry runs simultaneously in both banks with the lower
bank of fifteen adders dictating the maximum delay. With this
design a large percentage of the total multiply time (approx. 45%)
occurs at this stage. This percentage can be reduced by introducing
parallel operation, carry save or carry anticipate techniques. The
details of this are included later as they involve a lengthy
discussion.

4.3 A MULTIPLIER USING LOGARITHMS TO BASE FOUR

The principle of this design is exactly the same as the base
eight multiplier. The difference is the number of bits of the
multiplier examined in one cycle. Two bits are examined and thus
six cycles are required for a twelve bit multiplier.

Tables 9 and 10 contain the base four logarithms. The Karnough
map analysis is easier in both logarithm convertors, (L and L').
A complete logic diagram is shown in figure 14 and a detail of one
path through the circuit in figure 15.

4.4 MULTIPLICATION OF NEGATIVE NUMBERS

The logarithmic multipliers just described can be thought to
operate as a 'table look-up' system. This can be appreciated by
considering them as 'black-boxes', where the variables are entered
and the product appears after some delay. These 'black-boxes' have
been designed to operate on numbers with a positive binary integer
format. They cannot operate on one's or two's complement or any
other format. This can easily be proved by taking some simple

examples.

30

To incorporate negative number multiplication it becomes
necessary to use sign magnitude representation. This requires an
exclusive - OR operation between the sign bits of the multiplicand
and multiplier. The result of this is the sign of the product.
This short operation can run parallel with the multiplication.

CHAPTER FIVE

31

TIMING AND PRACTICAL CONSIDERATIONS
OF THE MULTIPLIERS

This chapter documents comparison of total delays through the
two types of logarithmic multiplier described in the last chapter.
Before this, suitable logic components are stated and it is on
these that the delay times are based. Finally, methods of reducing
the overall delay are discussed.

5.1 SUITABLE COMPONENTS

The data for the components mentioned below is based on Texas
7400 logic series [22] .

Multiplicand register - three four-bit bistable latches
supplying output and complement output (type SN7475).

Multiplier register - no parallel in/out shift register
with complemented output manufactured by Texas. Delay
times based on a suitable register constructed out of
discrete logic gates and J/K flip-flops.

Logic gates of converters (L and L') - the propagation
delays of these converters are based on Schottky T.T.L.

Simple index adders - As mentioned earlier, M.S.I. T.T.L.
adders may be used. For the base four method: type SN7483
and base eight: type SN7483 (4 bit) and SN7482 (2 bit).

32

Partial product adders - Suitable combinations of four bit
and two bit binary adders (ripple carry) and carry-save-adders
(SN7483, SN7482, SN74H183 respectively.)

Partial product shift registers - fifteen bits of this
24-bit register have to be parallel in/out, right shift
type (two SN74198). The remainder should be of serial in
parallel out, right shift type (two SN7164).

Partial product temporary storage register - latches as in
multiplicand register (three SN7475).

5.2 COMPARISON OF BASE EIGHT AND BASE FOUR MULTIPLIERS

The main difference in the complexity of the two designs occurs
in the logarithm converters. To convert a twelve bit word to
simple index numbers requires 60 logic gates in the base eight
method and only 35 in the base four. To convert back to the
semi-partial product requires 256 gates in base eight and 78 in base
four design. This makes the base four design cheaper and easier
to construct.

The propagation delays for both circuits are shown in figure 16.
These are based on the maximum delays stated by Texas, and applied
to the longest path through each multiplier. The timings assume
that the multiplier and multiplicand registers have been previously

33

loaded.

When determining the total time for the multiplication the following points need to be considered.

1. The copy of the complete partial product into the latches after each cycle can be done parallel with the beginning of the next cycle.

2. The final cycle does not contain shift or latch copy operations.

The complete multiplication times are:

BASE 8 :   420 x 4 - 90 = 1590 ns.

BASE 4 :   360 x 6 - 60 = 2100 ns.

5.3 METHODS OF DECREASING THE TOTAL PROPAGATION DELAY

There are five methods of reducing the delay of the logarithmic circuits. They involve:

1. Parallel operation of sections of the circuit.

2. Carry anticipate

3. Carry store

4. Skip cycle if multiplier bits are zero

5. combination of 1 and 4 with 2 or 3.

These techniques can be applied to both circuits but will only be considered for the base eight multiplier.

1. Parallel operation of sections of the circuit.

This requires that the first bank of adders in the partial product add stage should be of the carry-save type.

The principle of the parallel operation is to divide the circuit

---

34

into two. This is done by storing the carry and sum from the carry-save-adders in latches and operating the logic before then in parallel with that after. The logic diagram can be seen in Figure 17 and the timing diagram in figure 18 contains the total delay of this method. There is only an approximate reduction of 130 ns with this method.

There are two main disadvantages of this circuit. Firstly, the control signals become more complex with two separate shift controls required. Secondly, the cost increases due to the use of carry-save-adders and more latches.

2. Carry Anticipate

This is discussed by Lewin [23] . The principle is to examine all the inputs to the adders and simultaneously generate the carries for each stage. These carrie s are then applied to the appropriate adder stage to give the final sum. The delay for such a circuit is eight gates for the sum outputs and seven for the final stage carry output. The carry anticipate is, as normal, considered here in groups of five adders.

A bank of carry-save-adders has to be used again and with Schottky T.T.L. the delay time for the partial product stage is:

Carry anticipate: 110 ns + 18 ns for C.S. Adders

M.S.I. ripple carry: 188 ns.

The time for a base eight multiplication reduces from 1590 ns to 1350 ns but involves a large increase in circuit complexity and expense.

Figure 19 gives the circuit for carry anticipate.

3. Carry Store

In this system the adders are all replaced by carry-save-adders. The carry from each stage is stored and added back two adders to the right after the complete partial product has been shifted right three bits. After the final stage the carries are added to the appropriate sums in a fourteen stage ripple carry adder. This principle is shown in figure 20. This again requires extra control signals and logic. The delay is now reduced to 1420 ns.

4. Skip cycle if multiplier bits zero

When the multiplier bits are zero it is a waste of time to allow for the propagation delay of one complete cycle. An OR-gate connected to the outputs of the multiplier bits to be used in the next cycle will indicate if the shift control is to be doubled. This would skip over the three bit groups of the multiplier when zero.

The reduction in delay will obviously depend on the numbers multiplied.

5. Combination of previous methods

The above methods can be combined and the delay time can be reduced to approximately 1250 ns. The complexity and cost become considerable when this is done and it is worth considering more conventional methods and comparing them with the logarithmic multiplier first.

Appendix D contains an account of a software simulation process for checking the proposed designs of logarithmic multipliers.

35

CHAPTER SIX

COMPARISON OF LOGARITHMIC MULTIPLIERS
WITH CONVENTIONAL MULTIPLIERS

The logarithmic multipliers described, differ in structure from other designs only in the way the partial product is derived. The methods described in the previous chapter for summing the complete partial products can be applied to the multipliers to be described here. Therefore in the comparison it is only necessary to consider the stages producing the complete partial product.

The logarithmic multipliers are suitable for sign magnitude multiplication and should therefore be compared only with other types using this negative number representation.

6.1 A TWO BIT SHIFT MULTIPLIER USING A CONVENTIONAL APPROACH

The multiplier described here can be compared with the base four design. It operates on addition alone and examines two bits of the multiplier in any one cycle. The appropriate action on examining two bits of the multiplier is given in table 11. A logic diagram of a multiplier using this principle is given in figure 21. The circuit is self-explanatory and it suffices to say that the shifting of the multiplicand to effect multiplication by two is accomplished

36

37

by interposing gates between the multiplicand and adder.

The propagation delay for this circuit can be reduced further by using carry anticipate or skip cycle techniques. This is not necessary for a comparison providing it is made with the base four multiplier in figure 14 as far as the first bank of adders.

The following are delay times through each multiplier to the complete partial product stage.

Delay through base four multiplier.

$$= L + \left[3 \text{ bit Adder } A_1 \rightarrow C_3 \right] + L^1 + 2 \times \left[4 \text{ bit adder } Co \rightarrow C_4 \right] \\ + \left[4 \text{ bit adder } Co \rightarrow \Sigma_3 \right]$$

$$= 10 + 60 + 15 + 96 + 60$$

$$= 241 \text{ ns}$$

Delay through conventional 2 bit shift multiplier

$$= 3 \text{ gates } + 2 \times \left[4 \text{ bit adder } Co \rightarrow C_4 \right] + \left[4 \text{ bit adder } Co \rightarrow \Sigma_3 \right]$$

$$= 15 + 48 \times 2 + 60 \text{ ns}$$

$$= 171 \text{ ns}$$

From these results and by inspecting the relevant logic diagrams, the conventional approach is preferable for cost and speed.

6.2 A THREE BIT SHIFT MULTIPLIER USING A CONVENTIONAL APPROACH

This is a similar design to the two bit multiplier described above. Three bits of the multiplier are examined in one cycle. The appropriate actions as regard to the multiplier bits are given in table 12 and the logic diagram in figure 22.

A comparison with the same considerations as in section 6.1 can be made (as far as partial product stage).

38

Delay through base eight multiplier

$$= L + \left[6 \text{ bit adder} \right] + L' + 2 \times \left[4 \text{ bit adder } Co \rightarrow C_4 \right] \\ + \left[6 \text{ bit adder } Co \rightarrow \Sigma_3 \right]$$

$$= 10 + \left[48 + 42 \right] + 20 + 2 \times 48 + 60$$

$$= 276 \text{ ns}.$$

Delay through conventional three bit multiplier

$$= 4 \text{ gates } + 1 \text{ C.S.A.} + 3 \times \left[4 \text{ bit adder } Co \rightarrow C_4 \right] + 1 \text{ C.S.A.}$$

$$= 20 + 18 + 144 + 18$$

$$= 200 \text{ ns}.$$

Again there is an appreciable reduction in propagation delay with the conventional approach.

6.3 CONCLUSION

It is obvious from the two preceding chapters that the conventional approach to sign magnitude multiplication is faster, less complicated and considerably cheaper than the proposed logarithmic method.

SUMMARY

No more information can be extracted from Ludgate's 1909 report.
As this is the only information available about the machine [1] ,
our appraisal of his design must be limited to the ingenuity of
the theory. The detail of Ludgate's drawings, which largely
determines the machine's feasibility, cannot be determined.
Therefore, we must agree with a review in 1909 of Ludgate's
report by Boys [25] that,

"Until more detail as to the proposed construction and
drawings are available it is not possible to form any
opinion as to the practicability or utility of the
machine as a whole."

It can clearly be seen by reading this report, that the base
four and base eight multipliers have no advantages to offer over
conventional designs. Multipliers written to bases higher than
eight (i.e. 16, 32 ...) could be designed, but with difficulty due
to their complexity. They would also still suffer the same disadvantages as the base four and eight designs.

One further application of this type of logarithm is in the
design of a binary coded decimal multiplier. This is discussed
further in appendix B. It appears from a superficial design
consideration, that a logarithmic b.c.d. multiplier may have
advantages over existing types. This aspect deserves further
consideration.

39

This logarithmic method of multiplication may be applied to
software table look-up multipliers. Here the restrictions on using
logarithms to a low base number to reduce the circuit complexity
are removed, although other restrictions will be imposed. It may
be possible to apply logarithms to a b.c.d. table look-up multiplier.
These aspects are also worthy of further consideration.

Other aspects of Ludgate's engine have since been used in later
machines. Any further consideration of these aspects should be with
these later computers where more detail is available.

Other than the logarithms and their applications, the analytical
engine requires no further study unless new information on the machine
is found.

40

41

R E F E R E N C E S

1. Randell, B. "Ludgate's Analytical Machine of 1909"
   University of Newcastle upon Tyne, Computing
   Laboratory, Technical Report Series, no. 15. (1971)

2. Babbage, C. "Passages From The Life of A Philosopher."
   London, Longman 1864

3. Bowden, B.V. "Faster Than Thought". Pitman (1953)

4. Chase, G.C. "History of Mechanical Computing Machinery."
   Proceedings of the A.C.M. National Meeting,
   Pittsburgh 2 - 3 May 1952.

5. Hartree, D.R. "Calculating Instruments and Machines."
   Cambridge, 1950.

6. Encyclopaedia Britannica 13th ED. "Calculating Machines"

7. Ludgate, P.E. "Automatic Calculating Machines".
   Napier Tercentenary Celebration : Handbook of the
   Exhibition. Edinburgh : Royal Society of Edinburgh

8. Babbage, H.P. "On the Mechanical Arrangements of the
   Analytical Engine of the late Charles Babbage."
   Report of the Brit. Assoc. for the Advancement of
   Science (1888).

9. Morrison, P. and Morrison, E. "Charles Babbage and
   his Calculating Engines : Selected Writings by
   Charles Babbage and Others", Dover Publications (1961)

10. Wilkes, M.V. "Automatic Digital Computers". London,
    Methuen 1956.

11. Ludgate, P.E. "On a Proposed Analytical Machine"
    Scientific Proceedings, Royal Dublin Society, April 1909

12. Burks, A.W., Goldstine, H.H., and von Neumann, J.
    "Preliminary discussion of the logical design of
    an electronic computing instrument." Report to U.S.
    Army Ordnance Department, reprint in Bell and Navell [15] .

13. Boys, C.V. "A New Analytical Engine". Nature 81,2070
    1st July 1909.

14. Kilburn, T., Edwards, D.B.G., Langian, M. and Sumner, F.,
    "One-level Storage System". IRE Trans EC-11 pp
    223-235, April 1962.

15. Bell, C.G. and Newall, A. "Computer Structures: Readings
    and Examples". McGraw-Hill (1971)

16. Buchholz, W. "Planning a Computer System." McGraw-Hill
    (1962)

17. Flynn, M.J. "Very High Speed Computing Systems".
    IEEE Proc. Vol. 54 Dec' 66 pp 1901-1909

18. Wallace, C.S. "A Suggestion for a Fast Multiplier."
    IEEE Trans. on Electronic Computers, Vol. EC-13,
    No. 1 Feb. 1964.

19. Ferrari, D. "A Division Method Using a Parallel Multiplier."
    IEEE Trans. on Electronic Computers, Vol. EC-16, No. 2,
    April 1967 pp. 224-226.

20. Robertson, J.E. "A New Class of Digital Division Methods."
    IRE Trans. on Electronic Computers Vol. EC-17 Sept 1968
    pp 218-222

21. Anderson, S.F., Earle, J., Goldschmidt, R.E. and Powers, D.M.
    "The IBM System/360 - Model 91 : Floating Point Execution
    Unit". IBM Journal of R. and D., Vol. 11, No. 1, Jan 1967
    pp 48-53.

22. Texas Instruments Semiconductor Components - Digital
    Integrated Circuits, July 1971, Issue 2.

23. Lewin, D. "Theory and Design of Digital Computers." Nelson
    (1972).

24. Hollingdale, S.H. and Toatill C.C. "Electronic Computers"
    Pelican (1970).

25. Boys, C.V. "A New Analytical Engine." Nature 81,2070,
    1st July 1909, pp 14-15.

26. Ahmad, M. "Iterative Schemes for High Speed Division." The
    Computer Journal, Vol. 15, No. 4, Nov. 1972.

27. "D.A. 70 User Manual", Automation Division, Poole. (software
    simulation of logic circuits).

42

APPENDIX 'A'

INDEX NUMBER SYSTEMS

43

Definition of A UNIT :

> A positive integer in the range $[0, x-1]$,
>
> where x is the base to which the number system
>
> is written.

An Algorithm to derive:- index numbers of any number system.

1. Decide which base to write logarithms in.

2. Determine Prime numbers of the units of that number system.

3. Associate the simple index number 0 with unit 1.

4. Associate next unused index number with next highest prime number.

5. Calculate index numbers of all possible products of units so far produced (including units produced in this stage.)

6. Go to 7 if any product has two different index numbers associated with it or an index number has two separate products. Else go to 8 .

7. Erase last step 5 and associate next unused index number with last prime number. Go to 5 .

8. If any prime numbers left then go to Step 4 .

9. Associate lowest unused index number with zero.

10. Calculate index numbers of all possible products of units x zero.

11. If any index number associated with two products then go to 12, else go to 13 .

12. Erase lasts steps 9 and 10 and associate next unused index number with zero. Go to 10 .

44

13. END.

An example of the above algorithm is:

UNIT	INDEX NUMBER
1	Logs. to be written to Base 10.
2	Prime numbers are 1, 2, 3, 5, 7.
3	Simple index of 1 is 0

1	0

4   Simple index of 2 is 1

2	1

5   Index numbers of products that are powers of 2

4	2
8	3
16	4
32	5
64	6

128 not required since $> 9 \times 9 = 81$

4   3 has the next highest index number

3	7

5   Calculate the index numbers of all possible products of units defined so far.

9	14
27	21
81	28
6	8
12	9
18	15
24	10
36	16
48	11
54	22
72	17

45

UNIT	INDEX NUMBER

Unit 4 already has an index number

4	Let 5 have the next highest unused index number
5	12

5	Calculate index of Powers of 5
25	24

5	Calculate products of 5 and previous unit variables
10	13
15	8

7   Error here due to 8 already being an index number ∴ the index of 5 is changed. Same problem occurs at index 13, 18, 14, 20. ∴ 5 has index 23

5	23

5   Calculate index numbers that are products of previously defined units

25	46
10	24
20	25
40	26
15	30
45	37
30	31

By letting the index of 7 be 27, 29, or 32 the same problem as above occurs
Let 7 have index 33

7	33
14	34
28	35
56	36
21	40
42	41
63	47
35	56

Associate unit zero with highest available index. This, in this case eventually becomes 50.

46

Theory behind the derivation of the index numbers.

The above algorithm determines the lowest set of index numbers for any particular base. The principle of the algorithm is to associate the lower index numbers with the prime numbers or units that occur more often in forming all possible products of units.

An exception to the rule is the unit zero. It is convenient to give this the largest index number. Then all compound index numbers above a certain value may be regarded as zero.

The unit 'one', has to have a simple index of zero in every case (i.e. $\log z + \log 1 = \log z$)

Where two prime numbers are divisible into the same number of products, then their index numbers can be interchanged without increasing the index range. One example of this is the interchange of five and seven in Ludgate's logarithms to base ten.

APPENDIX B

A BINARY CODED DECIMAL HARDWARE
MULTIPLIER

The intention of this section is to briefly consider and describe how the previous theory of logarithms can be applied to a binary coded decimal multiplier.

The normal approaches to binary coded decimal (B.C.D.) multiplication involve software table look-up methods. The programming approach to multiplication is inexpensive to include in a B.C.D. machine, although it does require storage locations for its 'working' and tables. Another disadvantage is that it is slow, although with this type of computer high speed is not normally necessary.

Conventional methods of hardware multiplication are not viable with a B.C.D. number format. This is due to it being necessary to examine the multiplier in four bits and also because the multiplicand cannot be shifted, then added, to allow for multiplication. Hardware conversion from B.C.D. numbers to ordinary binary and back again involves repeated multiplication and division, respectively, by ten [24] . This is a costly and lengthy process and makes conversion of the numbers to a form suitable for an ordinary multiplier, impractical.

A logarithmic multiplier has the advantage in that it examines the multiplicand in groups of bits, and yet as a whole word as in conventional designs. The remainder of the logarithmic design can

be based on the same principle as the base four and base eight designs described in the body of the report. The final stage will have to use B.C.D. adders to add the semi-partial products. This is a disadvantage due to B.C.D. adders being complicated and expensive. [23] .

It appears from this short account that a logarithmic B.C.D. multiplier may be feasible, if a fast multiplier is required.

APPENDIX C

LOGIC SYMBOLS

49

1. AND GATE

A
B   Z = A.B.C
C

2. NAND GATE

A
B   Z = $\overline{A.B.C}$
C

3. INVERTOR

A ——▷o— Z = $\bar{A}$

4. OR GATE

A
B   Z = A + B

5. BISTABLE LATCH (STORE)

A_{t-1}

⊠ —— accept

A_t

6. CARRY SAVE ADDER

A B C

carry — CSA

Σ

7. BANK OF
FULL ADDERS

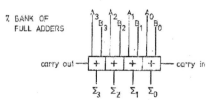

A_3 B_3 A_2 B_2 A_1 B_1 A_0 B_0

carry out — + + + + — carry in

Σ_3 Σ_2 Σ_1 Σ_0

8. PARALLEL IN/OUT REGISTER

parallel inputs

copy —

parallel
outputs

9. PARALLEL IN/OUT SHIFT REGISTER

parallel inputs

copy
shift

parallel
outputs.

10. BINARY NUMBER TO SIMPLE INDEX CONVERTER.

50

Ⓛ
*

11. COMPOUND INDEX NUMBER TO SEMI—PARTIAL
PRODUCT CONVERTER.

Ⓛ'
*

* { inputs to upper semi-circle,
    outputs from lower semi-circle.

APPENDIX D

PROGRAM SIMULATION OF LOGARITHMIC MULTIPLIERS

51

This appendix contains an account of the procedure used to check the logic design of the logarithmic converter stages (L and L'), using the DA70 simulation program [27]

Only one path through the multiplier need be simulated. The logic diagram is converted or coded to a form that is acceptable by the program; coding is explained in the 'DA70 USER MANUAL'. The simulation instructions are then written. Figure 23 contains suitable instructions for checking the accuracy of the logarithmic design.

Firstly all inputs permanently low are set to zero (all unconnected pins automatically set to logic 1). The multiplier and multiplicand registers are defined as one register, (INP). Taking the base eight design as an example, a six bit register is declared with its three least significant bits acting as the inputs for the multiplier and the others for the multiplicand.

The register, INP, is set to zero and then repeatedly incremented until it has covered its complete range. Thus all possible inputs are simulated. After each increment the input pulses are allowed to ripple through the simulated logic. The product enters another register, (OUT). The contents of the two registers are then printed before the next increment. Using the printed information the design can be checked.

It is also possible to use the DA70 simulation suite to produce accurate timing diagrams. This is explained in the User Manual and only a brief account will be made here.

52

A modified version of the 'simulation commands' in fig. 23 is sufficient and this is given on page 54. The commands can be explained as follows;

#DLAY - enables the propagation and edge times (from logic zero to one and one to zero) to be defined by the programmer. The actual delays are based on Texas components [22].

#PRUD - is used to set the propagation times of the adders. The alphabetic characters following each #PRUD refer to the pin connections to the adders. The delay is set in the connections which is effectively the same as being in the adder.

#DREG and #SREG - define the input and output registers and set them to zero initially.

The remainder of the program is basically the same as that described on page 51. The following commands are inserted

#MREG #SLDE and #RATE 4 - these produce the timing diagram. The format of the diagram is

+ equivalent to logic one

/ " " indeterminate state

- " " logic zero

The RATE 4 causes a symbol (+/or -) to be printed every fourth time-slot or four nanoseconds. (The average over the four nanoseconds is printed).

Examples of the timing diagrams for the base eight multiplier design are given on pages 55-60. Note that the corresponding final values of the input (INP) and output (OUT) registers are given after each timing diagram.

53

Due to an omission in the simulation commands fourteen of the sixty four timing diagrams produced by this program are inaccurate. For these fourteen, 32 nanoseconds have to be added to the total indicated delay times.* These are not included in the examples, except for 5.

The maximum indicated delay occurs at 5 x 2

$$= 23 \text{ timeslots} + 32\text{ns error}$$

$$= 92 + 32$$

$$= 124\text{ns}.$$

This test does not give the exact maximum delay. The delay is dependent on the initial conditions of the logic (from previous multiplication). Since the program takes into account the difference in propagation times from logic one to zero and zero to one, the delay is dependent upon the initial conditions of the logic, (i.e. the previous multiplication.)

The maximum delay could be found by further programming, but this could not be over 135ns. This compares favourably with the estimated delay which does not account for edge times

$$\text{estimated delay} = L + (6 \text{ bit adder } C_0 \longrightarrow \Sigma_4) + L^1$$

$$= 10 + (48 + 42) + 20$$

$$= 120\text{ns}.$$

*The error in the simulation is the omission of a # PRUD command to set the delay of the carry between the four stage adder and the two stage adder to 27, 35. (i.e. 27ns delay for logic 0→1 and 35ns for logic 1→0.) A nominal 4, 3 is automatically assumed, thus an allowance of 32ns is necessary on multiplications where a carry occurs between the adders.

54

```
JA70        DATE 10/04/73
#SIML SIML 1
#PROG
#DLAY 1PAND 5 5 1 1
#DLAY 2NAND 5 5 1 1
#DLAY 3NAND 5 5 1 1
#DLAY 4NAND 5 5 1 1
#DLAY 8NAND 5 5 1 1
#PRUD 1=55 50 H=29 28 G=19
#PRUD F=14 15 H1=35 30 NR=1
#PRUD HG=19 19 HF=14 15
****

#PRUD HF=35 30 HD=29 28
#PRUD D=29 28 F=35 30
#PRUD G=5 5 F=11 7 N=21 25
#PRUD M=26 29 L=5 5 K=11 7
#EXT AV=0
#DREG INP=A B C HA HB HC
#SREG INP=0
#DREG OUT=R S T U V L
#STRT
#MREG INP OUT
#GIPO
#RATE 6
RPT #RINC INP
#RUN 200
#PREG
#IFREG INP=0 END
#GOTO RPT
END #END
```

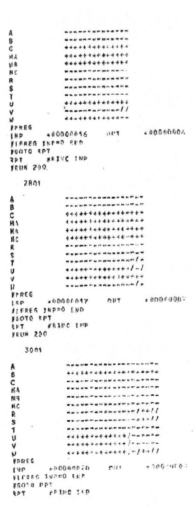

55

```
#IFREG INP=0 END
#GOTO RPT
RPT      #RINC INP
#RUN 200

   11801

A        +++++++++++++++++++
B        +++++++++++++++++++  ┐7
C        +++++++++++++++++++  ┘
MA       ++++++++++++++++++
MB       --------------------  ┐4
MC       ---------------------  ┘
R        ------------/--/
S        +++++++++++++++++  ┐3
T        ++++++++++++++/++  ┘
U        +++++++++++++//++
V        ++++++++++++//--  ┐4
W        ++++++++++++++--/--  ┘
#PREG
INP   +00005074    OUT   +00060003
#IFREG INP=0 END
#GOTO RPT
RPT      #RINC INP
#RUN 200
```

                                    input   output (result)

Example of timing diagram
produced by simulation program.

7 × 4 = 34
        └octal

```
A        --------------------
B        --------------------
C        +++++++++++++++++++
MA       +++++++++++++++++++
MB       +++++++++++++++++++
MC       --------------------
R        --------------------
S        --------------------
T        --------------------
U        ++++++++++++++++
V        -------------//
W        ++++++++++++++
#PREG
INP   +00000016    OUT   +00060004
#IFREG INP=0 END
#GOTO RPT
RPT      #RINC INP
#RUN 200

   2801

A        --------------------
B        --------------------
C        +++++++++++++++++
MA       ++++++++++++++++++
MB       +++++++++++++++++++
MC       +++++++++++++++++++
R        --------------------
S        --------------------
T        ----------------/-
U        ++++++++++++++/-/
V        +++++++++++++++++
W        --------------/-
#PREG
INP   +00000017    OUT   +00060007
#IFREG INP=0 END
#GOTO RPT
RPT      #RINC INP
#RUN 200

   3001

A        --------------------
B        ++++++++++++++++++
C        --------------------
MA       --------------------
MB       --------------------
MC       ----------------
R        --------------/++//
S        --------------+//
T        --------------
U        ++++++++++++/---
V        ++++++++++++/---
W        +++++++++++=/+//
#PREG
INP   +00000020    OUT   +00060005
#IFREG INP=0 END
#GOTO RPT
RPT      #RINC INP
```

56

An analysis of Ludgates machine leading to a design of a digital logarithmic multiplier

D. Riches, Department of Electrical and Electronic Engineering, University College, Swansea, U.K., 1973. © 1973, University College Swansea, reproduced with permission

57

58

D. Riches, Department of Electrical and Electronic Engineering, University College, Swansea, U.K., 1973. © 1973, University College Swansea, reproduced with permission

An analysis of Ludgates machine leading to a design of a digital logarithmic multiplier

#RUN 200

11201

```
A          ++++++++++++++++
B          +++++++++++++--+++
C          +++++++++++++--+++
MA         ----------------
MB         ----------------
MC         +++++++++++++--++
R          -----------------//--
S          ---------------/+//
T          --------------//--
U          -------------////
V          -----------------/+++
W          ------------------////
```
#PREG
INP   *00000071   OUT   *00000007
#IFREG INP=0 END
#GOTO RPT
RPT   #RINC INP
#RUN 200

11401

```
A          +++++++++++++++++++
B          +++++++++++++++++++
C          +++++++++++++++++++
MA         -------------------
MB         +++++++++++--1+---+++
MC         --------------------
R          -----------------//
S          ------------------
T          ------------------//
U          +++++++++++++++--++
V          +++++++++++++++--++
W          +++++++++++++++++//
```
#PREG
INP   *00000072   OUT   *00000010
#IFREG INP=0 END
#GOTO RPT
RPT   #RINC INP
#RUN 200

11601

```
A          ++++++++++++++++++
B          ++++++++++++++++++++
C          ++++++++++++++++++++
MA         --------------------
MB         ++++++++++++++++++++
MC         ++++++++++++++++++++
R          --------------------
S          -----------------//
T          ++++++++++++/---/+//
U          ++++++++++++/---/++
V          ++++++++++++/--/+//
W          -----------------/++
```
#PREG
INP   *00000073   OUT   *00000020

8201

```
A          +++++++++++++++++++++++
B          -----------------------
C          +++++++++++++++++++++++
MA         -----------------------
MB         +++++++++++++++++++++++
MC         -----------------------
R          -----------------/--------
S          -----------------/----/+--
T          --------------/---/++++++
U          ++++++++++++++++//+---
V          --------------------/----/
W          +++++++++++++++++---------
```
PREG
NP   *00000052   OUT   *00000012
IFREG INP=0 END
GOTO RPT
PT   #RINC INP
RUN 200

6401

. . . .

Maximum delay occurs when
multiplying   5 × 2.

$= 23 \text{ timeslots} \times 4 + 32\text{ns error}$

$= 92 - 32 \text{ ns}$

$= 124 \text{ ns}$

N.B.   This diagram is
incorrect.   see *, page 53

61

TABLES
nos. 1–12

62

TABLE 1 - LUDGATE'S SIMPLE INDEX NUMBERS

63

TABLE 2 – LUDGATE'S COMPOUND INDEX NUMBERS

COMPOUND INDEX NO.	PARTIAL PRODUCT	COMPOUND INDEX NO.	PARTIAL PRODUCT
0	1	33	7
1	2	34	14
2	4	35	28
3	8	36	56
4	16	37	45
5	32	38	-
6	64	39	-
7	3	40	21
8	6	41	42
9	12	42	-
10	24	43	-
11	48	44	-
12	-	45	-
13	9	46	25
14	18	47	63
15	36	48	-
16	72	49	-
17	-	50	0
18	-	51	0
19	-	52	0
20	-	53	0
21	27	54	-
22	54	55	-
23	5	56	35
24	10	57	0
25	20	58	0
26	40	59	-
27	-	60	-
28	81	61	-
29	-	62	-
30	15	63	-
31	30	64	0
32	-	65	-
		66	49

TABLE 2

64

TABLE 3 – MULTIPLICATION EXAMPLE (728 x 35)

65

Decimal Digit Represented by rod	Protrusion Distance of Rod [1 - 10 units]
0	1
1	10
2	9
3	6
4	8
5	3
6	5
7	2
8	7
9	4

TABLE 4

66

TABLE 5 - POSSIBLE INSTRUCTIONS FOR LUDGATE'S MACHINE

Store Address (96 locations in each store, 7 bit address)

1. Rs ....store address (rotate store to position s)

Store Format

1. $B_O := S_O$ ....outer store variable to buffer
2. $B_I := S_I$ ....inner store variable to buffer
3. B_O^I S_O^I
   B_I := S_I ....both store variables to buffers.

Data automatically re-written before accessing new location,
          (i.e. shuttles replaced before store rotated)

Operation code

1. $A := 0$ ....clear accumulation
2. $A := A + B_Q \times 1$ ....add buffer contents to accumulator
3. $A := A - B_Q \times 1$ ....subtract buffer contents from accumulator
4. $A := A + B_O \times B_I$.......add product of buffers to accumulator
5. $A := A + B_O \div B_I^1$ ....add division of outer by inner buffers to
                                 accumulator
6. $B_Q^1 := A$ ....store accumulator in buffer(s).
7. Sk ....formula-paper conditional skip.

Input { 8. $B_{Q1}^1 := K$ ....keyboard data to buffer(s)
        9. $B_Q^1 := P$ ....number-paper data to buffer(s)

Output { 10. Po := A ....output accumulator via Punch.
         11. Pr := A ....output accumulator via Printer.

12. Io := Bo ....temporary store index register from outer
                                 buffer.
13. $A := A + B_I \times Io$ ....add product of index register to inner buffer
14. $T := B_Q$ ....temporary store buffer
15. $B_Q := T$ ....load buffer from temporary store.

where Q = 1 for inner store   instructions 12 - 15 allow:
         0 for outer store
                                 (a) direct multiplication of any two
      Q^1 = 1 or 0 or both   variables in different stores.

* assumed $\frac{Bo}{B_I}$ (b) direct transfer of data from one
                                     store location to another.

TABLE 5

UNIT TO BASE 8	SIMPLE INDEX NO.
0	45
1	0
2	1
3	6
4	2
5	17
6	7
7	22

TABLE 6 - LOGARITHMS TO BASE 8

Compound Index Number	Semi-Partial Product	Compound Index Number	Semi-Partial Product
0	1	27	-
1	2	30	25
2	4	31	52
3	10	32	-
4	20	33	-
5	40	34	-
6	3	35	-
7	6	36	31
10	14	37	-
11	30	40	-
12	0	41	43
13	-	42	-
14	11	43	-
15	22	44	61
16	44		
17	5	45	0
20	12	46	0
21	24	47	0
22	7	53	0
23	16	54	0
24	34	64	0
25	17	67	0
26	35	112	0

T A B L E   7

COMPOUND INDEX NUMBERS OF

BASE EIGHT LOGARITHMS

69

TABLE 8 - KARNAUGH MAPS FOR L' CONVERTER
(Assuming logic zero for most significant input)

70

UNIT TO BASE 4				SIMPLE INDEX NO.		
Binary	Decimal	Base 4		Base 4	Decimal	Binary
0 0	0	0		13	7	1 1 1
0 1	1	1		0	0	0 0 0
1 0	2	2		1	1	0 0 1
1 1	3	3		3	3	0 1 1
P Q						A B C

TABLE 9 - BASE FOUR, SIMPLE INDEX NUMBERS

COMPOUND INDEX NUMBER				PARTIAL PRODUCT		
Binary [R S T U]	Decimal	Base 4		Base 4	Decimal	Binary [V W X Y]
0 0 0 0	0	0		1	1	0 0 0 1
0 0 0 1	1	1		2	2	0 0 1 0
0 0 1 0	2	2		10	4	0 1 0 0
0 0 1 1	3	3		3	3	0 0 1 1
0 1 0 0	4	10		12	6	0 1 1 0
0 1 1 0	6	12		21	9	1 0 0 1
0 1 1 1	7	14		0	0	0 0 0 0
1 0 0 0	8	20		0	0	0 0 0 0
1 0 1 0	10	22		0	0	0 0 0 0
1 1 1 0	14	3		0	0	0 0 0 0

TABLE 10 - BASE FOUR, INDEX TO PARTIAL PRODUCT

$V = \bar{R}.S.T.\bar{U}$

$U = S.\bar{T} \vee \bar{R}.\bar{S}.T.U$

$X = \bar{R}.\bar{S}.U \vee S.\bar{T}$

$Y = \bar{R}.S.T.\bar{U} \vee \bar{R}.\bar{S}.T.U \vee \bar{R}.\bar{S}.\bar{T}.\bar{U}$

Boolean equations converting outputs of L' converter to its inputs.

An analysis of Ludgates machine leading to a design of a digital logarithmic multiplier

71

MULTIPLIER BITS	ACTION
00	Do nothing
01	Add multiplicand
10	Add multiplicand shifted left one place (i.e. x 2)
11	Add multiplicand, and add multiplicand shifted left one place.

TABLE 11 - TWO BIT MULTIPLIER ACTIONS

MULTIPLIER BITS	ACTION
000	Do nothing
001	Add MPCD
010	Add MPCD shifted left once
011	Add MPCD shifted left once, Add MPCD.
100	Add MPCD shifted left twice
101	Add MPCD shifted left twice, Add MPCD.
110	Add MPCD shifted left twice, Add MPCD shifted left once
111	Add MPCD shifted left twice, Add MPCD shifted left once and Add MPCD

MPCD = multiplicand

TABLE 12 - THREE BIT MULTIPLIER ACTIONS

72

FIGURES
nos. 1—23

73

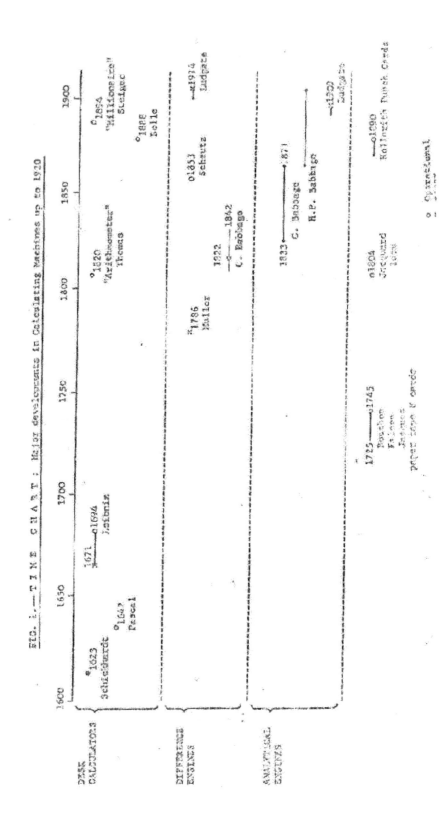

FIG. 1. — T I M E   C H A R T : Major developments in Calculating Machines up to 1920

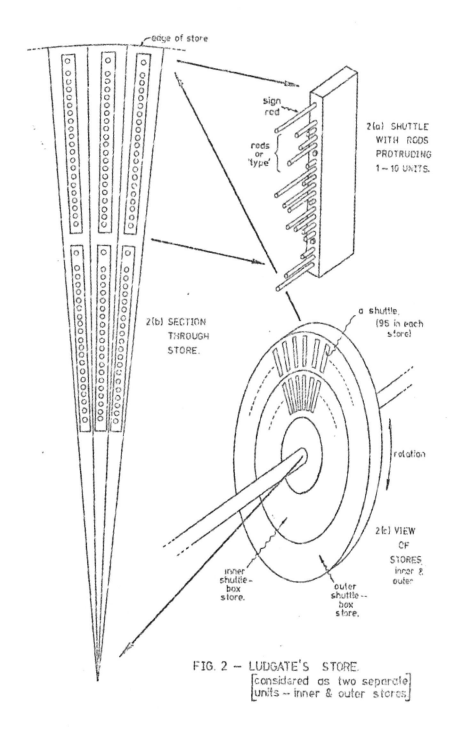

74

2(a) SHUTTLE WITH RODS PROTRUDING 1 – 10 UNITS.

sign rod

rods or 'type'

edge of store

2(b) SECTION THROUGH STORE.

a shuttle (95 in each store)

rotation

2(c) VIEW OF STORES inner & outer

inner shuttle-box store.

outer shuttle-box store.

FIG. 2 — LUDGATE'S STORE.
[considered as two separate units – inner & outer stores]

An analysis of Ludgates machine leading to a design of a digital logarithmic multiplier

D. Riches, Department of Electrical and Electronic Engineering, University College, Swansea, U.K., 1973. © 1973, University College Swansea, reproduced with permission

75

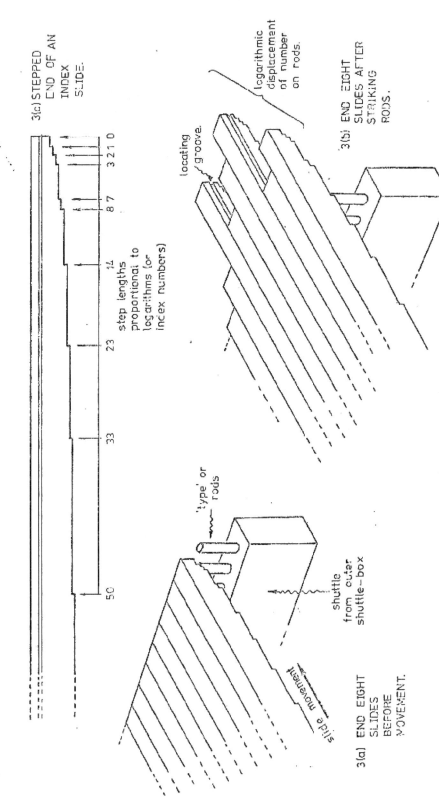

3(c) STEPPED END OF AN INDEX SLIDE.

step lengths proportional to logarithms (or index numbers)

logarithmic displacement of number on rods.

3(b) END EIGHT SLIDES AFTER STRIKING RODS.

locating groove.

'type' or rods

shuttle from outer shuttle-box

slide movement

3(a) END EIGHT SLIDES BEFORE MOVEMENT.

FIG. 3 — STEPPED ENDS OF THE INDEX SLIDES.

76

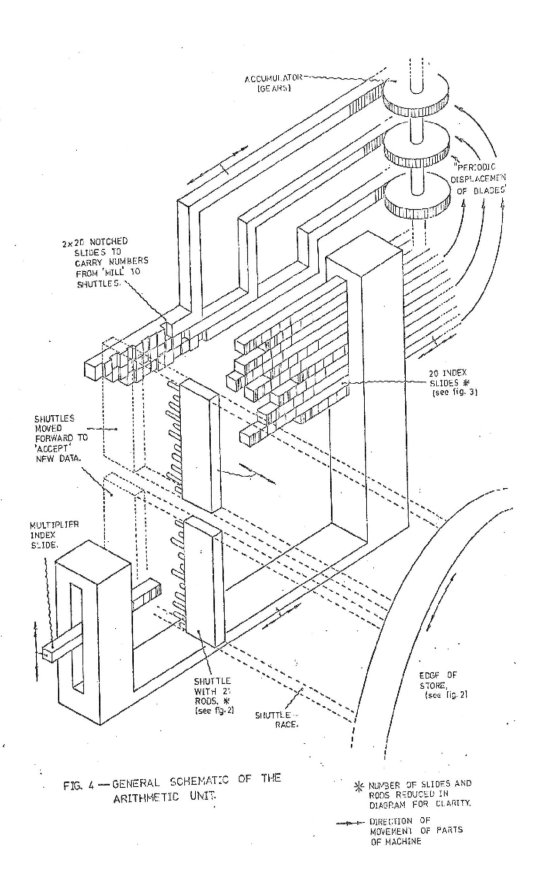

ACCUMULATOR
(GEARS)

"PERIODIC
DISPLACEMENT
OF BLADES"

2×20 NOTCHED
SLIDES TO
CARRY NUMBERS
FROM 'MILL' TO
SHUTTLES.

20 INDEX
SLIDES *
(see fig. 3)

SHUTTLES
MOVED
FORWARD TO
'ACCEPT'
NEW DATA.

MULTIPLIER
INDEX
SLIDE.

EDGE OF
STORE,
(see fig. 2)

SHUTTLE
WITH 20
RODS, *
(see fig. 2)

SHUTTLE
RACE.

FIG. 4 — GENERAL SCHEMATIC OF THE
ARITHMETIC UNIT.

* NUMBER OF SLIDES AND
RODS REDUCED IN
DIAGRAM FOR CLARITY.

——► DIRECTION OF
MOVEMENT OF PARTS
OF MACHINE

D. Riches, Department of Electrical and Electronic Engineering, University College, Swansea, U.K., 1973. © 1973, University College Swansea, reproduced with permission

An analysis of Ludgates machine leading to a design of a digital logarithmic multiplier

77

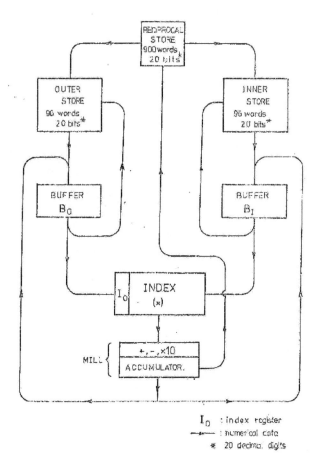

FIG-5.—BASIC BLOCK DIAGRAM
OF LUDGATE'S MACHINE.

78

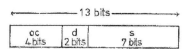

INSTRUCTION WORD FORMAT
(derived from table 5.)

oc - operation code

d - store format inner, outer
or both

s - store address

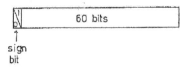

DATA WORD FORMAT

FIG. 6

79

Explanation of Figure -. 7

Both operands taken from same address, one from each store, and placed in buffers (B_o and B_I) and accumulator cleared. Multiplication proceeds and result forms in the accumulator. The branches in the flow chart describe the six possible ways of storing the result. The following numbered comments correspond to the diagram.

1. Result written over operand in inner store

   i.e. operand in inner store lost

   operand in outer store retained.

2. Result written over operand in outer store

   i.e. operand in outer store lost

   operand in inner store retained.

3. Result written over both operands (outer and inner)

   i.e. both operands lost

4. Rotate store, then 5, 6 or 7 as below.

   i.e. both operands retained

5. Store result in outer store.

6. Store result in inner store.

7. Store result in inner and outer stores (having same address).

80

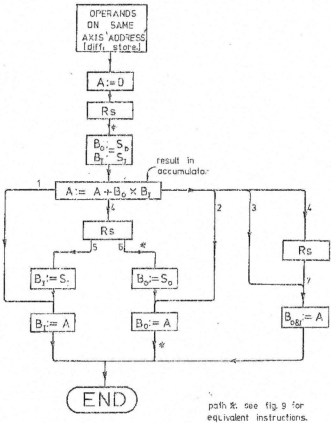

FIG. 7 -- MULTIPLICATION FLOW CHART.

81

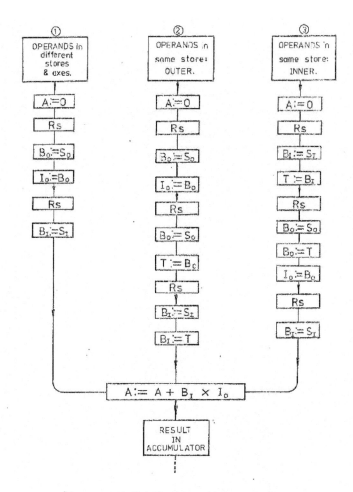

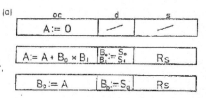

FIG. 8 – MULTIPLICATION FLOW CHART.

82

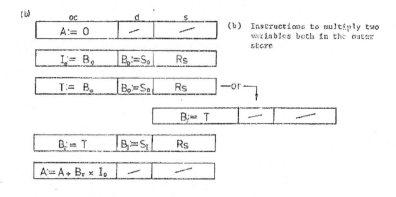

(a) Instructions to multiply two variables on same axis, result stored in another location in outer store [e.g. Fig. 9 Path?]

(b) Instructions to multiply two variables both in the outer store

(c) Instructions to add two variables to the accumulator, result in the accumulator

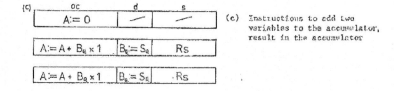

FIG. 9 – EXAMPLES OF INSTRUCTION WORDS

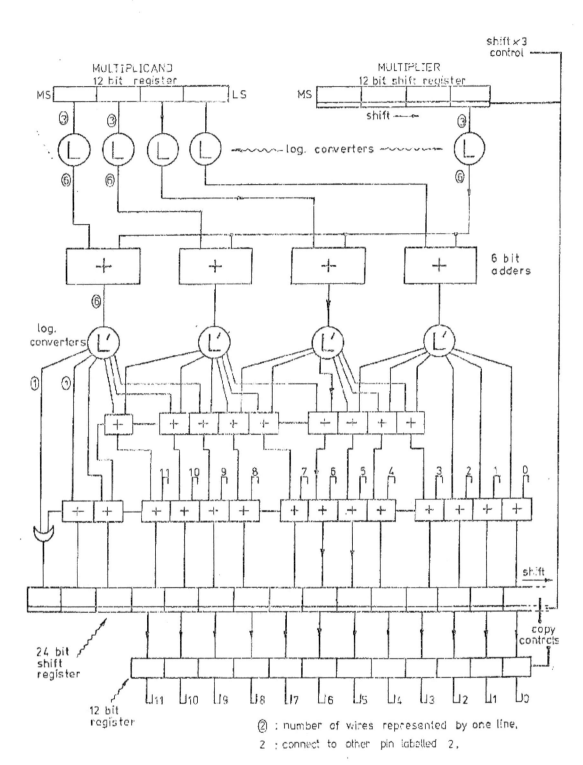

83

shift x3
control

MULTIPLICAND
12 bit register
MS   LS

MULTIPLIER
12 bit shift register
MS

shift

log. converters

6 bit
adders

log.
converters

11 10 9 8   7 6 5 4   3 2 1 0

shift

24 bit
shift
register

copy
controls

12 bit
register

L11 L10 L9 L8 L7 L6 L5 L4 L3 L2 L1 L0

② : number of wires represented by one line.

2 : connect to other pin labelled 2.

FIG. 10 — LOGIC DIAGRAM OF
BASE EIGHT MULTIPLIER.

An analysis of Ludgates machine leading to a design of a digital logarithmic multiplier

D. Riches, Department of Electrical and Electronic Engineering, University College, Swansea, U.K., 1973. © 1973, University College Swansea, reproduced with permission

84

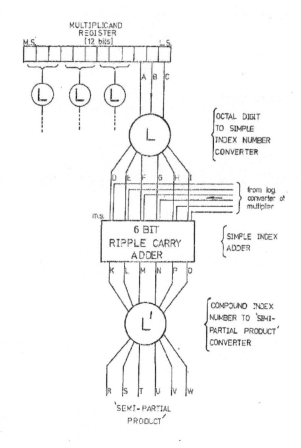

FIG. 11 –
ONE PATH THROUGH MULTIPLIER.
[BASE EIGHT DESIGN]

85

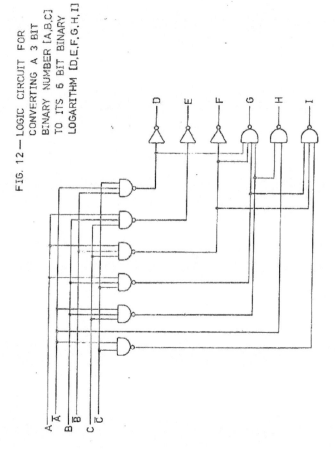

FIG. 12 – LOGIC CIRCUIT FOR
CONVERTING A 3 BIT
BINARY NUMBER [A,B,C]
TO ITS 6 BIT BINARY
LOGARITHM [D,E,F,G,H,I]

86

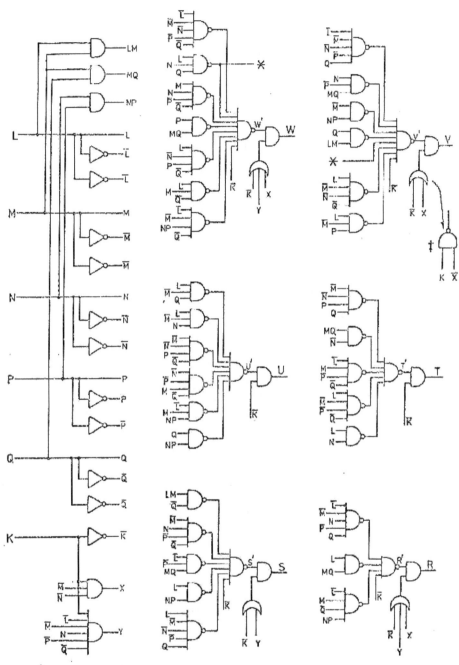

where 5 6 or 7 input gate
occurs, assume 8 input gate
with remaining inputs high.

† OR gates shown for clarity,
may be replaced by nand
gates.

FIG. 13—LOGIC DIAGRAM FOR
COMPOUND INDEX NUMBER
[K,L,M,N,P,Q] TO SEMI-
PARTIAL PRODUCT [R,S,T,U,V,W]
CONVERTER.

87

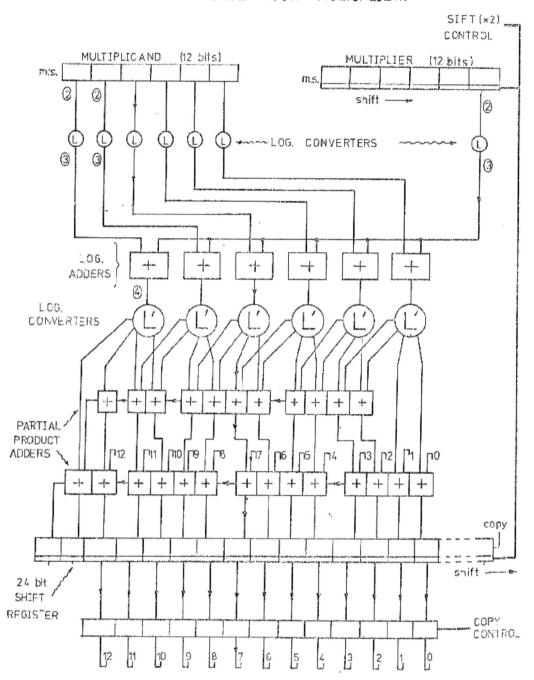

FIG. 14 — LOGIC DIAGRAM OF BASE FOUR MULTIPLIER.

② : number of signals represented by one line,

2 : connect to other pin labelled 2,

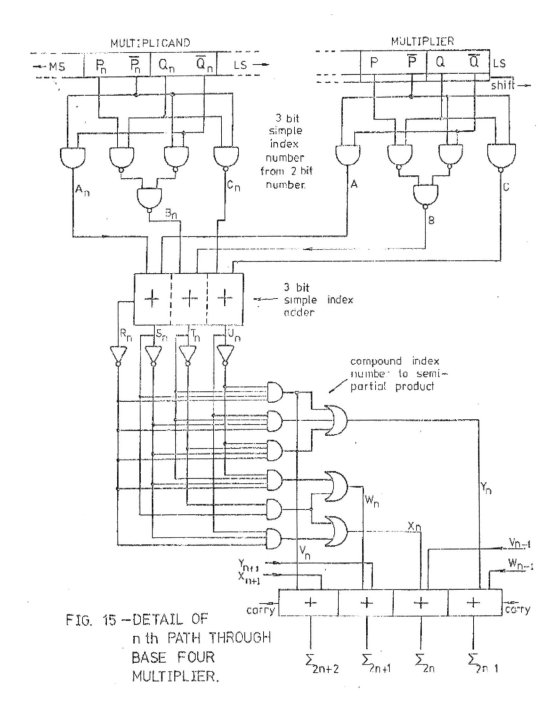

FIG. 15 – DETAIL OF
n th PATH THROUGH
BASE FOUR
MULTIPLIER.

88

89

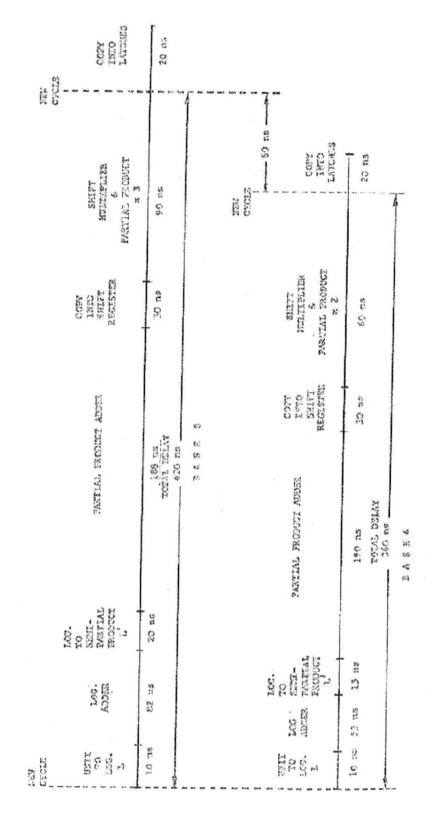

FIG. 16 — TIMING DIAGRAMS OF MULTIPLIERS FOR ONE CYCLE
(AS DESCRIBED IN FIGURES 2.2 & 2.6)

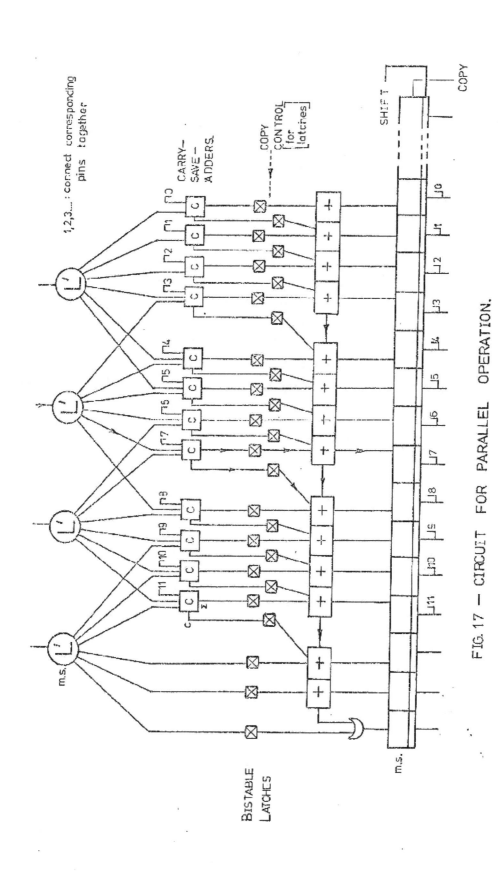

FIG. 17 — CIRCUIT FOR PARALLEL OPERATION.

An analysis of Ludgates machine leading to a design of a digital logarithmic multiplier

90

91

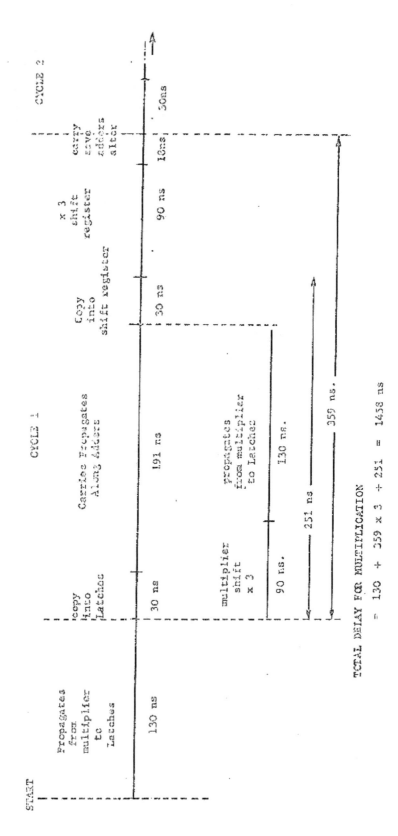

TOTAL DELAY FOR MULTIPLICATION

$= 130 + 359 \times 3 + 251 = 1458$ ns

FIG. 18 — TIMING DIAGRAM FOR PARALLEL OPERATION OF BASE 8 MULTIPLIER

92

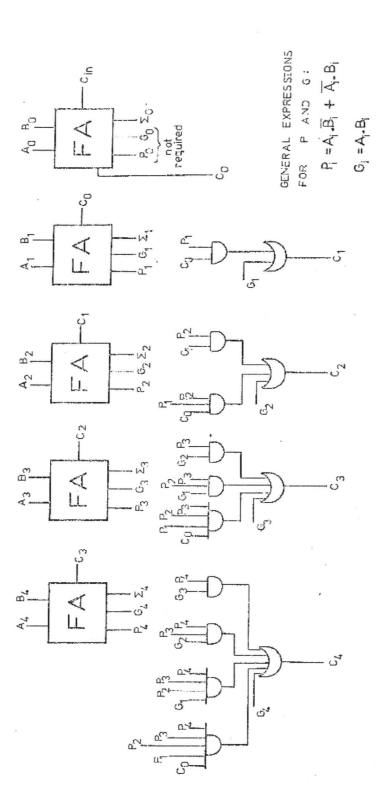

GENERAL EXPRESSIONS
FOR P AND G :

$$P_i = A_i . \overline{B_i} + \overline{A_i} . B_i$$

$$G_i = A_i . B_i$$

FIG. 19 — FIVE STAGE CARRY LOOK-AHEAD ADDER.

An analysis of Ludgates machine leading to a design of a digital logarithmic multiplier

93

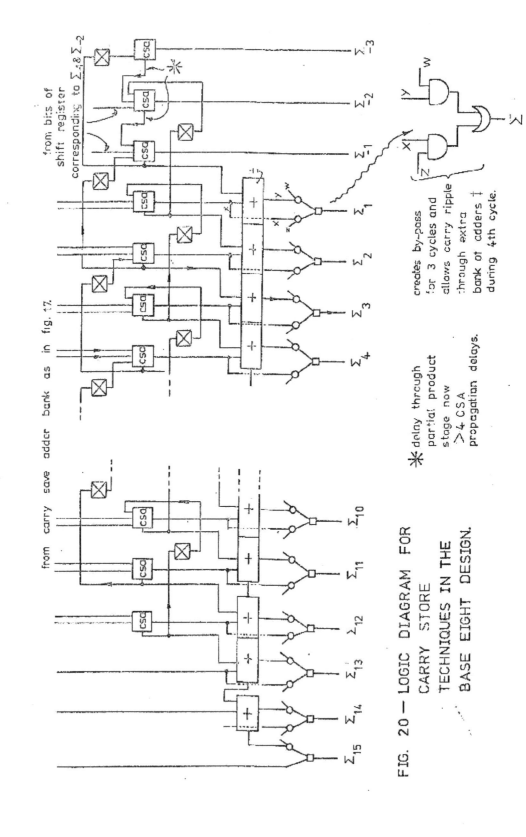

FIG. 20 — LOGIC DIAGRAM FOR
CARRY STORE
TECHNIQUES IN THE
BASE EIGHT DESIGN.

* delay through
partial product
stage now
> 4 CSA
propagation delays.

* creates by-pass
for 3 cycles and
allows carry ripple
through extra
bank of adders ‡
during 4th cycle.

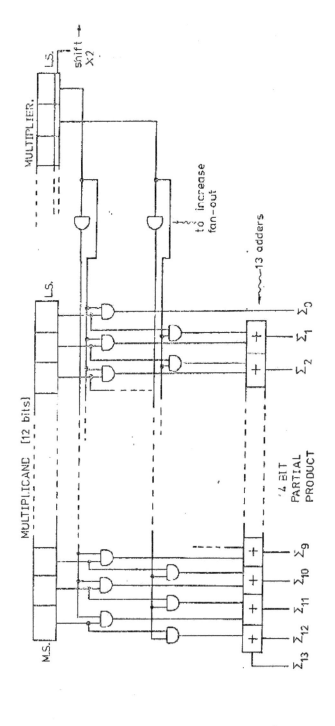

FIG. 21 — TWO BIT SHIFT MULTIPLIER.
[operating on addition alone]

An analysis of Ludgates machine leading to a design of a digital logarithmic multiplier

D. Riches, Department of Electrical and Electronic Engineering, University College, Swansea, U.K., 1973. © 1973, University College Swansea, reproduced with permission

95

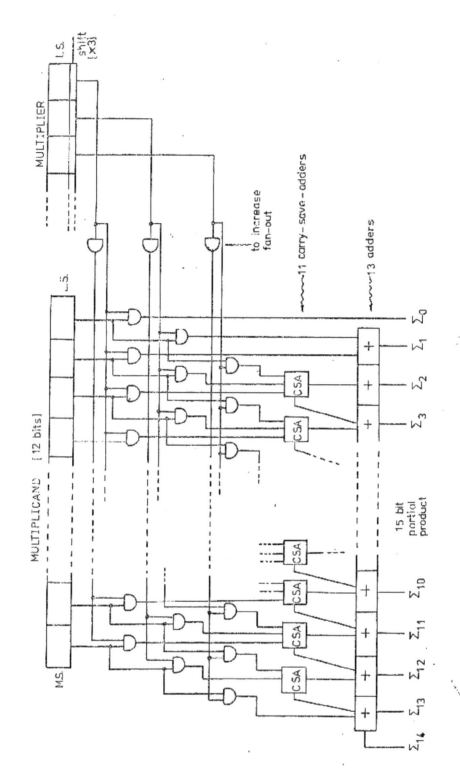

FIG. 22 — THREE BIT SHIFT MULTIPLIER.
[operating on addition alone]

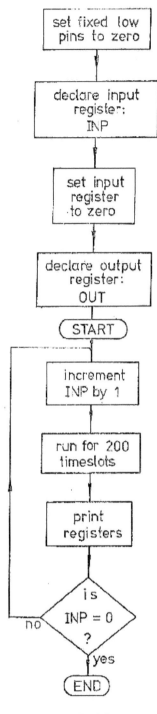

FIG. 23
SIMULATION PROGRAM
FLOW CHART.

96

An analysis of Ludgates machine leading to a design of a digital logarithmic multiplier

An analysis of Ludgates machine leading to a design of a digital logarithmic multiplier

D. Riches, Department of Electrical and Electronic Engineering, University College, Swansea, U.K., 1973.

Ludgate's analytical machine of 1909

B. Randell

Computing Laboratory, Claremont Tower, University of Newcastle upon Tyne

This paper discusses the little known analytical machine, or program-controlled mechanical calculator, designed by Percy E. Ludgate in Ireland during the years 1903 to 1909, and documents the results of a search for information about his life and work.

(Received January 1971)

1. Introduction

Whilst seeking information about Charles Babbage and Ada Augusta, Countess of Lovelace, for use as introductory material for a lecture on the evolution of programming, I came across a reference to a paper entitled 'Automatic Calculating Engines' by one Percy E. Ludgate (1914). The paper proved to contain a brief but competent account of Babbage's Analytical Engine, but, to my great surprise, it ended with the following comments:

'I have myself designed an analytical machine, on different lines from Babbage's, to work with 192 variables of 20 figures each. A short account of it appeared in the *Scientific Proceedings, Royal Dublin Society*, April 1909. Complete descriptive drawings of the machine exist, as well as a description in manuscript, but I have not been able to take any steps to have it constructed.'

I immediately sought out a copy of this 1909 paper and found that it consisted of a fascinating account of a machine which was indeed quite comparable to, yet quite different from, Babbage's famous Analytical Engine. At the time I had never questioned the more or less standard account, found in the introductions to many textbooks on computing, which jump straight from the work of Babbage, started in 1834, to that of Aiken (and sometimes Stibitz and Zuse) just over 100 years later.* The purpose of the present paper is to put Ludgate's work into perspective, and to document the results of an extensive search for further information about his life and work. His 1909 paper and a review of it by Professor C. V. Boys that appeared shortly afterwards are reprinted in full as Appendices 1 and 2.

2. Background

Charles Babbage died in 1872, having failed to complete either the full scale version of his difference engine, or his analytical engine (a large amount of original material on Babbage's engines has been reprinted by Morrison and Morrison, 1961). Shortly before his death part of the analytical engine (the arithmetic mill, together with a printing device) was put together and is now in the Science Museum, London (see Baxandall, 1926). In 1878, a committee of the British Association for the Advancement of Science wrote a report (Merrifield, 1878) which praised the basic conception of the analytical engine 'as a marvel of mechanical ingenuity and resource', but concluded that 'in the present state of the design it is not more than a theoretical possibility; that is to say, we do not consider it a certainty that it could be constructed and

put together so as to run smoothly and correctly, and to do the work expected of it. . . . we have come, not without reluctance, to the conclusion, that we cannot advise the British Association to take any steps, either by way of recommendation or otherwise, to procure the construction of Mr. Babbage's Analytical Engine'. However, according to Babbage's son (H. P. Babbage, 1888) an arithmetic unit (presumably the one referred to above), was used to give a successful demonstration to the committee of the operation of addition, with parallel assimilation of the carry digits.

Shortly afterwards Babbage's son put together another arithmetic unit and printing device, using pieces of mechanism and designs left by his father. This was completed and in 1910 by way of a demonstration was used to calculate and print successive multiples of π (Baxandall, 1926). This machine is now also in the Science Museum. In contrast, it would seem that the other major components of the engine, namely the store (originally planned to have a storage capacity for 1,000 numbers, each of 50 decimal digits, later reduced to 200 numbers, each of 25 digits) and the control mechanism were never constructed.

It is, in retrospect, clear that the complete analytical engine was far ahead of the technology of the time—indeed it has been claimed that Babbage's efforts were worthwhile merely for the benefits that they brought to mechanical engineering.

The boldness of Babbage's plans becomes clear when one realises that it was only during the mid-nineteenth century that a calculating machine (the arithometer of Thomas de Colmar, the first version of which was invented in 1820) achieved commercial success. In fact it was only towards the end of the century that mechanical calculating machines received widespread use (see Wilkes, 1956). On the other hand, Jacquard looms, whose technique of punched card control Babbage intended to use, were well established quite early in the nineteenth century. However, the use of punched cards for recording logical and numerical data had to await the work of Hollerith in the 1880s. His system, which was electro-mechanical, was used with great success in the 1890 US Census and within a few years had spread to several European countries, although for several years card handling was manual, rather than mechanised.

One can, therefore, with hindsight, claim that by the turn of the century the time had become much more propitious for the development of an analytical engine, or as we would now term it, a program-controlled computer. However, Ludgate's contribution was not that of making a second attempt to implement Babbage's machine, taking advantage of the improved technological capabilities of the day. Rather, he claims that

---

*I have since found just two accounts of the history of computers which even mention Ludgate's description of his analytical machine, namely Hoffman (1962) and Wilkes (1956). The latter contains the most detailed modern appraisal of Babbage's analytical engine that I have encountered.

Brian Randell, The Computer Journal, Volume 14, Issue 3, pp. 317–326, 1971. © 1971, British Computer Society, reproduced with permission

Ludgate's Analytical Machine of 1909

Brian Randell, The Computer Journal, Volume 14, Issue 3, pp. 317–326, 1971. © 1971, British Computer Society, reproduced with permission

until the later stages of his efforts, he had been in ignorance of Babbage's work, and his design is sufficiently novel for this to be accepted. Indeed all three main components of his analytical machine, the store, the arithmetic unit and the sequencing mechanism show evidence of considerable ingenuity and originality.

3. The store

The method of data storage that Ludgate designed used a 'shuttle' for each variable. Each shuttle acted as a carrier for a set of protruding metal rods, there being one rod for the sign, and for each of the 20 decimal digits comprising a number. The current value of each digit of the number currently stored in the shuttle was represented by the lateral position of the corresponding rod, i.e. by the length of rod protruding from the shuttle. The shuttles were to be held in 'two co-axial cylindrical shuttle-boxes'. A particular number could be brought to the arithmetic unit by rotating the appropriate shuttle box through an appropriate angle. There was also to be provision for tables of constants, represented by sets of holes, of depth from one to nine units, drilled into the surface of one or more special cylinders.

Assuming that there are appropriate means of transferring data between this type of representation and that used in the arithmetic unit (a topic on which Ludgate's paper is rather obscure), this method of storage would appear to have considerable advantages over that used by Babbage, i.e. columns of toothed discs, each capable of being connected by a train of gear wheels to the arithmetic unit. Certainly it is very convenient to access a number merely by an appropriate rotation of a cylindrical shuttle box. Ludgate mentions a further advantage, i.e. 'that the shuttles are quite independent of the machine, so that new shuttles, representing new variables can be introduced at any time'—one could perhaps claim that this was the forerunner of the modern replaceable disk!

4. The arithmetic unit

It is in the arithmetic unit that Ludgate's machine differs most markedly from that of Babbage, and indeed, as far as I can prove, from all other mechanical calculating machines. The unit is a 'direct' or 'partial product' multiplying machine, rather than one in which multiplication is performed by repeated addition. Direct multiplying machines already existed by the time of Ludgate. The first successful one was that of Bollée invented in 1889, although patents had earlier been granted to Barbour in 1872 and to Verea in 1878 for machines of this type. Indeed by the turn of the century a direct multiplying machine, known as the 'Millionaire', was starting to achieve wide distribution in Europe and America (see Chase 1952).

Each of these machines performed their multiplication of individual digits from the two different operands by what was in essence a table-look-up on a complete multiplication table. (Bollée represented the table by an array of 100 pairs of rods, each rod being one to nine units long—there is no means of knowing whether this was the inspiration for Ludgate's method of number storage.)

In Ludgate's machine what is essentially a logarithmic method of multiplication is used. Each digit of one operand is translated into the corresponding 'index number' (or 'Irish logarithm', as Boys so delightfully terms it). This set of index numbers is then added to the index number form of one of the digits of the other operand. The additions are performed con-

currently by simple concatenation of lateral displacements. Then a reverse translation is performed to obtain the set of two-digit partial products. (The description of the mechanism for doing all this is somewhat obscure, and gives one a clearer impression of its ingenuity than its practicality.) The set of partial products so obtained for each digit in the second operand are then accumulated using a 'mill', which is presumably a fairly conventional set of co-axial toothed wheels incorporating a carrying mechanism. Ludgate claims that he designed his own version of Babbage's 'anticipating carriage', i.e. mechanism for assimilation, in a single step, of all the carry digits produced during the addition of two numbers (described in Babbage, 1851), but gives no details of his design.

Ludgate was equally unconventional in his scheme for division, which instead of using repeated subtractions was based upon a table of reciprocals of the integers 100 to 999, and a rapidly convergent series for $(1 + x)^{-1}$, where $|x| < 10^{-3}$. The calculation of the series was controlled by what we would now call a built-in subroutine.

5. The sequencing mechanism

The sequencing mechanism that Ludgate describes has more in common with that used on the Harvard Mk. 1 (Aiken and Hopper, 1946) nearly 40 years later, than that designed by Babbage for his analytical engine.* Ludgate's machine was to be controlled by a perforated paper tape, termed a 'formula paper', on which each row of perforations defined a complete instruction. Each instruction specified two operands, the type of arithmetic operation to be performed, and the location (or pair of locations) which was to receive the result. Babbage on the other hand, for some unknown reason, intended to use two distinct sets of Jacquard cards, one for specifying which variables were to provide the operands for and receive the result from, each operation (the so-called 'variable cards'), the other for specifying the sequence of types of operations ('operation cards'). Furthermore, there were to be means for economising on operation cards (but not apparently variable cards) by indicating the number of times that there were to be repeated applications of the same type of arithmetic operation, rather than supply a sequence of identical operation cards (see Lady Lovelace's translation of Menabrea's article (Menabrea, 1843 note D)). It is not clear how Babbage intended to use the specification, on a variable card, of a particular variable, to access the column of disks representing that variable; in Ludgate's machine, as mentioned earlier, all that was necessary was to arrange for the appropriate angle of rotation of the shuttle-box containing the shuttle representing the required variable.

Ludgate clearly agreed with Babbage as to the fundamental importance of conditional branching, although he does not indicate how it was to be done—presumably, following Babbage, he intended that the mechanism that read the formula paper could be directed to skip a specified number of rows, either forwards or backwards. (It is interesting to note that the original Harvard Mk. 1 had only a very limited form of conditional branching.)

A third feature of the sequencing mechanism was the provision of built-in subroutines. The operation code for division, for example, caused control to pass temporarily to a sequence of instructions represented by rows of perforations on a permanent 'dividing cylinder'. Another cylinder provided a logarithm subroutine, and Ludgate mentions the possibility of indefinite expansion of the set of such auxiliary cylinders.

---

*Interestingly enough, a memorandum written by Aiken (1937) outlining his plans for an automatic calculating machine, mentions Ludgate in addition to describing Babbage's work on difference and analytical engines. However, the reference, whose wording closely follows that used earlier by Baxandall (1926), merely lists Ludgate amongst the designers of difference engines, so there is little reason to suppose that Aiken was familiar with Ludgate's plans.

6. Percy E. Ludgate

The two papers by Ludgate and the review of the first of these by C. V. Boys, gave only a few meagre starting points for a search for further information about his life and work. Baxandall (1926) had indicated that Ludgate was Irish but it was not known whether this was based merely on the fact that his first paper had appeared in the Scientific Proceedings of the Royal Dublin Society.

A search of standard reference works proved fruitless, and no further papers by Ludgate were traced, leading to the surmise (later proved correct) that he had died at a fairly early age. Inquiries of academic institutions and societies, mainly in Ireland, were similarly unsuccessful. Eventually his niece, Miss Violet Ludgate, who luckily is still living in Dublin, was traced through the heroic efforts of Mr. Desmond Clarke, Librarian and Secretary of the Royal Dublin Society, who contacted each of the Ludgates listed in the Dublin telephone directory. The following details of Percy Ludgate's life were obtained either directly from Miss Ludgate, or by following up leads that she furnished.

Percy Edwin Ludgate was born on 2 August 1883, at the house of his parents Michael and Mary Ludgate in Townsend Street, Skibbereen, County Cork, Ireland. He was the youngest of four children, all boys, his brothers being named Thomas, Frederick and Alfred. His father, Michael Ludgate, was born at Mallow, County Cork, and was married whilst serving in the army. He and his wife spent a part of their married life in India, where their first child, Thomas, was born. The second child, Frederick, was born in Winchester in 1879. Later the family moved to Ireland, first to Skibbereen, and later to Dublin, where Percy was brought up. It is believed that Percy Ludgate attended North Strand Parish School, and that he studied accountancy at Rathmines College of Commerce, Dublin, and was awarded a gold medal by the Corporation of Accountants on the occasion of his final examination, which he passed with distinction. (Efforts to confirm these details of his education have not so far met with any success.) He attended St. George's Church, Temple Street, Dublin.

Percy Ludgate worked as an auditor until his death, with the firm of Kevans and Son, 31 Dame Street, Dublin, which later transferred to Westmoreland Street, and is now part of the firm of Cooper Brothers. It seems almost certain that his work on the analytical machine was a private hobby which, according to his niece, 'he used to work at nightly, until the small hours of the morning'. He never married. Quoting from another letter that I received from Miss Ludgate: 'Percy liked walking; he took long solitary walks. I do not think he had many other interests. He attended his parish church services regularly. He was very gentle, a modest simple man. I never heard him make a condemning remark about anyone. I would say he was a really good man, highly thought of by anyone who knew him. Always appeared to be thinking deeply.' The photograph (Fig. 1), is believed to have been taken a few years before his death.

The one other person I have traced who has recollections of Percy Ludgate is Mr. E. Dunne, of Cooper Brothers, who joined the firm of Kevans and Son early in 1921. According to Mr. Dunne, 'My association with Mr. Ludgate was quite brief, but I had known him by repute for some time . . . As a person he possessed the characteristics one usually associates with genius, and he was so regarded by his colleagues on the staff . . . Like all men of his stature he was humble, courteous, patient and popular, and his early death closed a career that was full of promise for the future . . . The books and other memoranda of his disappeared and whether they were taken away by Percy before he became ill, or treated as part of the flotsam when Kevans and Son moved to Westmoreland Street, I cannot unfortunately say.' During the 1914-18 war he worked for a committee, set up by the War Office, headed by Mr. T.

Fig. 1. Percy E. Ludgate

Condren-Flinn, senior partner of Kevans and Son. The task of this committee was to control the production and sale of oats, over a wide area of the country, in order to maintain a supply for the cavalry divisions of the army. This involved planning and organisation on a vast scale and Ludgate was much praised for the major role he played. It is interesting to note that this provides a further parallel to the work of Charles Babbage who, because of his book *On the Economy of Machinery and Manufactures*, has often been called one of the originators of what is now known as 'Operational Research'.

It has not proved possible to obtain any information about his contacts with Professor A. W. Conway, of University College Dublin, who communicated Ludgate's paper to the Royal Dublin Society. Similarly unsuccessful have been efforts to trace the present whereabouts of the papers of Professor C. V. Boys, in the hope of finding his correspondence with Ludgate. (The obituary notice for Professor Boys which appeared in the *Proceedings of the Physical Society* in November 1944 stated that his papers 'were found well preserved and in meticulous order at his death'.) Furthermore, the records of the committee set up by the Royal Society of Edinburgh to organise the Napier Tercentenary Celebration, for whose handbook Ludgate contributed the article entitled 'Automatic Calculating Engines', have apparently not been preserved. (For some unknown reason Ludgate is not included in the listing of names and affiliations of contributors given at the back of the handbook.) Finally, no records have been found of any attempts to patent the analytical machine, or to obtain financial backing for its construction from the government.

At his death, on 16 October 1922, which occurred shortly after his return from a holiday in Lucerne, and which was announced in a brief obituary notice in the *Irish Times* two days later, Percy Ludgate was living with his widowed mother and his brother Alfred, at 30 Dargle Road, Drumcondra, Dublin. He had developed pneumonia, and his brother Frederick's wife (or rather, widow, since Frederick had died nine months earlier) who had helped to nurse Percy during his

Brian Randell, The Computer Journal, Volume 14, Issue 3, pp. 317–326, 1971. © 1971, British Computer Society, reproduced with permission

Ludgate's Analytical Machine of 1909

fatal illness, contracted pneumonia herself and died six days after Percy, leaving a daughter, Violet, who is now the sole surviving descendant of Michael and Mary Ludgate.

In his will, drawn up some five years before his death, Percy Ludgate had appointed his brother Alfred as his executor, and had willed the residue of his estate to his mother. His assets, mostly government stocks, amounted to somewhat over £800, and included a mere £10 for his personal effects. There is no means of knowing whether his drawings and manuscripts relating to the analytical engine were amongst these personal effects. His mother died in 1946, aged 97, and his brother Thomas, who had lived most of his life in Peacehaven, Sussex, in 1951. If any drawings or manuscripts had remained in the family they would presumably have passed into the possession of Percy's brother Alfred. However, there is no indication that this happened, and at Alfred's death in 1953 no such papers were found amongst his effects.

7. Concluding remarks

It seems unlikely that Ludgate ever attempted to construct the machine described in his 1909 paper. In fact in the 1914 paper he implies that he had discarded the plans, in favour of a second design:

'The most pleasing characteristic of a difference engine made on Babbage's principle is the simplicity of its action, the difference being added together in unvarying sequence; but notwithstanding its simple action, its structure is complicated by a large amount of adding mechanism—a complete set of adding wheels with carrying gear being required for the tabular number, and every order of difference except the highest order. On the other hand, while the best feature of the analytical engine or machine is the Jacquard apparatus (which, without being itself complicated, may be made a powerful instrument for interpreting mathematical formula), its weakness lies in the diversity of movements the Jacquard apparatus must control. Impressed by these facts, and with the desirability of reducing the expense of construction, I designed a second machine in which were combined the best principles of both the analytical and difference types, and from which are excluded their more expensive characteristics. By using a Jacquard I found it possible to eliminate the redundancy of parts hitherto found in difference-engines, while retaining the native symmetry of structure and harmony of action of machines of that class. My second machine, of which the design is on the point of completion, will contain but *one* set of adding wheels, and its movements will have a rhythm resembling that of the Jacquard loom itself. It is primarily intended to be used as a difference-machine, the number of orders of differences being sixteen. Moreover, the machine will also have the power of automatically evaluating a wide range of miscellaneous formulae.'

Excepting the possibility that further searches, perhaps stimulated by this paper, succeed in locating Ludgate's designs or correspondence, or trace some hitherto unsuspected collaborator, our appraisal of him will have to remain based on the fragmentary evidence afforded by his two published papers. One must, however, wonder just how much more he might have achieved if he had had but a modest fraction of the resources available to Babbage (to say nothing of Aiken!), and had not succumbed to pneumonia at such a tragically early age.

8. Acknowledgements

I have been aided in the search for information on the life and work of Ludgate by many people, only a few of whom is it possible to mention here. First and foremost I am indebted to Miss Violet Ludgate, who has provided much information about her uncle, and who has made extensive efforts herself to trace further possible sources of information. Other individuals to whom I wish to express my gratitude include Mr. Desmond Clarke, of the Royal Dublin Society, Mr. E. Dunne of Cooper Brothers, Wilton Place, Dublin, Miss M. C. Griffith, of the Public Record Office of Ireland, Mr. P. Henchy, Director of the National Library of Ireland, Mr. B. J. Lynch, of the Institute of Chartered Accountants in Ireland, Mrs. Ann MacDonald, until recently librarian at the Computing Laboratory of the University of Newcastle upon Tyne, Professor T. Murphy and Professor T. E. Nevin of University College Dublin, Mr. M. Woodger of the National Physical Laboratory, and Mr. H. Woolfe, of the Science Museum Library, London.

Ludgate's 1909 paper is reproduced by the kind permission of the Royal Dublin Society, and the review by C. V. Boys by kind permission of the Editor of *Nature*. The quotations from Ludgate's paper of 1914 are reproduced by kind permission of the Royal Society of Edinburgh.

Appendix 1

(Reprinted from Scientific Proceedings, Royal Dublin Society 12, 9 (1909) pp. 77-91.)

ON A PROPOSED ANALYTICAL MACHINE
by
Percy E. Ludgate
(Communicated by Professor A. W. Conway, M.A.)
[Read February 23. Ordered for Publication March 9. Published April 28, 1909.]

I purpose to give in this paper a short account of the result of about six years' work, undertaken by me with the object of designing machinery capable of performing calculations, however intricate or laborious, without the immediate guidance of the human intellect.

In the first place I desire to record my indebtedness to Professor C. V. Boys, F.R.S., for the assistance which I owe to his kindness in entering into correspondence with me on the matter to which this paper is devoted.

It would be difficult and very inadvisable to write on the present subject without referring to the remarkable work of Charles Babbage, who, having first invented two Difference Engines, subsequently (about eighty years ago) designed an Analytical Engine, which was shown to be at least a theoretical possibility; but unfortunately its construction had not proceeded far when its inventor died. Since Babbage's time his Analytical Engine seems to have been almost forgotten; and it is probable that no living person understands the details of its projected mechanism. My own knowledge of Babbage's Engines is slight, and for the most part limited to that of their mathematical principles.

The following definitions of an Analytical Engine, written by Babbage's contemporaries, describe its essential functions as viewed from different standpoints:

'A machine to give us the same control over the executive which we have hitherto only possessed over the legislative department of mathematics.'[*]

'The material expression of any indefinite function of any degree of generality and complexity, such as, for instance: $F(x, y, z, \log x, \sin y, \&c.)$, which is, it will be observed, a function of all other possible functions of any number of quantities.'[†]

[*]C. Babbage: 'Passages from the Life of a Philosopher', p. 129.
[†]R. Taylor's 'Scientific Memoirs', 1843, vol. iii., p. 691.

'An embodying of the science of operations constructed with peculiar reference to abstract number as the subject of those operations.'*

'A machine for weaving algebraical patterns.'†

These four statements show clearly that an Analytical Machine 'does not occupy common ground with mere "calculating machines". It holds a position wholly its own'.

In order to prevent misconception, I must state that my work was not based on Babbage's results—indeed, until after the completion of the first design of my machine, I had no knowledge of his prior efforts in the same direction. On the other hand, I have since been greatly assisted in the more advanced stages of the problem by, and have received valuable suggestions from, the writings of that accomplished scholar. There is in some respects a great resemblance between Babbage's Analytical Engine and the machine which I have designed—a resemblance which is not, in my opinion, due wholly to chance, but in a great measure to the nature of the investigations, which tend to lead to those conclusions on which the resemblance depends. This resemblance is almost entirely confined to the more general, abstract, or mathematical side of the question; while the contrast between the proposed structure of the two projected machines could scarcley be more marked.

It is unnecessary for me to prove the possibility of designing a machine capable of automatically solving all problems which can be solved by numbers. The principles on which an Analytical Machine may rest 'have been examined, admitted, recorded, and demonstrated'.‡ I would refer those who desire information thereon to the Countess of Lovelace's translation of an article on Babbage's Engine, which, together with copious notes by the translator, appears in R. Taylor's 'Scientific Memoirs', vol. iii.; to Babbage's own work, 'Passages from the Life of a Philosopher'; and to the Report of the British Association for the year 1878, p. 92. These papers furnish a complete demonstration that the whole of the developments and operations of analysis are capable of being executed by machinery.

Notwithstanding the complete and masterly treatment of the question to be found in the papers mentioned, it will be necessary for me briefly to outline the principles on which an Analytical Machine is based, in order that my subsequent remarks may be understood.

An Analytical Machine must have some means of storing the numerical data of the problem to be solved, and the figures produced at each successive step of the work (together with the proper algebraical signs); and, lastly, a means of recording the result or results. It must be capable of submitting any two of the numbers stored to the arithmetical operation of addition, subtraction, multiplication, or division. It must also be able to select from the numbers it contains the proper numbers to be operated on; to determine the nature of the operation to which they are to be submitted; and to dispose of the result of the operation, so that such result can be recalled by the machine and further operated on, should the terms of the problem require it. The sequence of operations, the numbers (considered as abstract quantities only) submitted to those operations, and the disposition of the result of each operation, depend upon the algebraical statement of the calculation on which the machine is engaged; while the magnitude of the numbers involved in the work varies with the numerical data of that particular case of the general formula which is in process of solution. The question therefore naturally arises as to how a machine can be made to follow a particular law of development as expressed by an algebraic formula. An eminently satis-

factory answer to that question (and one utilised by both Babbage and myself) is suggested by the Jacquard loom, in which interesting invention a system of perforated cards is used to direct the movements of the warp and weft threads, so as to produce in the woven material the pattern intended by the designer. It is not difficult to imagine that a similar arrangement of cards could be used in a mathematical machine to direct the weaving of numbers, as it were, into algebraic patterns, in which case the cards in question would constitute a kind of mathematical notation. It must be distinctly understood that, if a set of such cards were once prepared in accordance with a specified formula, it would possess all the generality of algebra, and include an infinite number of particular cases.

I have prepared many drawings of the machine and its parts; but it is not possible in a short paper to go into any detail as to the mechanism by means of which elaborate formulae can be evaluated, as the subject is necessarily extensive and somewhat complicated; and I must, therefore, confine myself to a superficial description, touching only points of particular interest or importance.

Babbage's Jacquard-system and mine differ considerably; for, while Babbage designed two sets of cards—one set to govern the operations, and the other set to select the numbers to be operated on—I use one sheet or roll of perforated paper (which, in principle, exactly corresponds to a set of Jacquard-cards) to perform both these functions in the order and manner necessary to solve the formula to which the particular paper is assigned. To such a paper I apply the term formula-paper. Each row of perforations across the formula-paper directs the machine in some definite step in the process of calculation—such as, for instance, a complete multiplication, including the selection of the numbers to be multiplied together. Of course a single formula-paper can be used for an indefinite number of calculations, provided that they are all of one type or kind (i.e. algebraically identical).

In referring to the numbers stored in the machine, the difficulty arises as to whether we refer to them as mere numbers in the restricted arithmetical sense, or as quantities, which, though always expressed in numerals, are capable of practically infinite variation. In the latter case they may be regarded as true mathematical variables. It was Babbage's custom (and one which I shall adopt) when referring to them in this sense to use the term 'Variable' (spelt with capital V), while applying the usual meanings to the words 'number' and 'variable'.

In my machine each Variable is stored in a separate shuttle, the individual figures of the Variable being represented by the relative positions of protruding metal rods or 'type', which each shuttle carries. There is one of these rods for every figure of the Variable, and one to indicate the sign of the Variable. Each rod protrudes a distance of from 1 to 10 units, according to the figure or sign which it is at the time representing. The shuttles are stored in two co-axial cylindrical shuttle-boxes, which are divided for the purpose into compartments parallel to their axis. The present design of the machine provides for the storage of 192 Variables of twenty figures each; but both the number of Variables and the number of figures in each Variable may, if desired, be greatly increased. It may be observed, too, that the shuttles are quite independent of the machine, so that new shuttles, representing new Variables, can be introduced at any time.

When two Variables are to be multiplied together, the corresponding shuttles are brought to a certain system of slides called the *index*, by means of which the machine computes the product. It is impossible precisely to describe the mechanism of the index without drawings; but it may be compared to a slide-rule on which the usual markings are replaced by moveable blades. The index is arranged so as to give several readings simultaneously. The numerical values of the readings are indi-

*loc. cit., p. 694.

†loc. cit., p. 696.

‡C. Babbage: 'Passages from the Life of a Philosopher', p. 450.

Brian Randell, The Computer Journal, Volume 14, Issue 3, pp. 317–326, 1971. © 1971, British Computer Society, reproduced with permission

Ludgate's Analytical Machine of 1909

Ludgate's Analytical Machine of 1909

Brian Randell, The Computer Journal, Volume 14, Issue 3, pp. 317–326, 1971. © 1971, British Computer Society, reproduced with permission

cated by periodic displacements of the blades mentioned, the duration of which displacements are recorded in units measured by the driving shaft on a train of wheels called the *mill*, which performs the carrying of tens, and indicates the final product. The product can be transferred from thence to any shuttle, or to two shuttles simultaneously, provided that they do not belong to the same shuttle-box. The act of inscribing a new value in a shuttle automatically cancels any previous value that the shuttle may have contained. The fundamental action of the machine may be said to be the multiplying together of the numbers contained in any two shuttles, and the inscribing of the product in one or two shuttles. It may be mentioned here that the fundamental process of Babbage's Engine was not multiplication but addition.

Though the index is analogous to the slide-rule, it is not divided logarithmically, but in accordance with certain *index numbers*, which, after some difficulty, I have arranged for the purpose. I originally intended to use the logarithmic method, but found that some of the resulting intervals were too large; while the fact that a logarithm of zero does not exist is, for my purpose, an additional disadvantage. The index numbers (which I believe to be the smallest whole numbers that will give the required results) are contained in the following tables:

Column 1 of **Table 1** contains zero and the nine digits, and column 2 of the same Table the corresponding *simple index numbers*. Column 1 of **Table 2** sets forth all *partial products* (a term applied to the product of any two units), while column 2 contains the corresponding *compound index numbers*. The relation between the index numbers is such that the sum of the simple index numbers of any two units is equal to the compound index number of their product. **Table 3** is really a re-arrangement of Table 2, the numbers 0 to 66 (representing 67 divisions on the index) being placed in column 1, and in column 2, opposite to each number in column 1 which is a compound index number, is placed the corresponding simple product.

Now, to take a very simple example, suppose the machine is supplied with a formula-paper designed to cause it to evaluate x for given values of a, b, c, and d, in the equation $ab + cd = x$, and suppose we wish to find the value of x in the particular case where $a = 9247$, $b = 8132$, $c = 21893$, and $d = 823$.

The four given numbers are first transferred to the machine by the key-board hereafter mentioned; and the formula-paper causes them to be inscribed in four shuttles. As the shuttles of the inner and outer co-axial shuttle-boxes are numbered consecutively, we may suppose the given values of a and c to be inscribed in the first and second shuttles respectively of the inner box, and of b and d in the first and second shuttles respectively of the outer box; but it is important to remember that it is a function of the formula-paper to select the shuttles to receive the Variables, as well as the shuttles to be operated on, so that (except under certain special circumstances, which

Table 1

UNIT	SIMPLE INDEX NO.	ORDINAL
0	50	9
1	0	0
2	1	1
3	7	4
4	2	2
5	23	7
6	8	5
7	33	8
8	3	3
9	14	6

Table 2

PARTIAL PRODUCT	COMP. INDEX NO.	PARTIAL PRODUCT	COMP. INDEX NO.	PARTIAL PRODUCT	COMP. INDEX NO.
1	0	15	30	36	16
2	1	16	4	40	26
3	7	18	15	42	41
4	2	20	25	45	37
5	23	21	40	48	11
6	8	24	10	49	66
7	33	25	46	54	22
8	3	27	21	56	36
9	14	28	35	63	47
10	24	30	31	64	6
12	9	32	5	72	17
14	34	35	56	81	28

Comp. index numbers of zero: 50, 51, 52, 53, 57, 58, 64, 73, 83, 100.

Table 3

COMP. INDEX NO.	PARTIAL PRODUCT	COMP. INDEX NO.	PARTIAL PRODUCT
0	1	34	14
1	2	35	28
2	4	36	56
3	8	37	45
4	16	38	—
5	32	39	—
6	64	40	21
7	3	41	42
8	6	42	—
9	12	43	—
10	24	44	—
11	48	45	—
12	—	46	25
13	—	47	63
14	9	48	—
15	18	49	—
16	36	50	0
17	72	51	0
18	—	52	0
19	—	53	0
20	—	54	—
21	27	55	—
22	54	56	35
23	5	57	0
24	10	58	0
25	20	59	—
26	40	60	—
27	—	61	—
28	81	62	—
29	—	63	—
30	15	64	0
31	30	65	—
32	—	66	49
33	7		

arise only in more complicated formulae) any given formula-paper always selects the same shuttles in the same sequence and manner, whatever be the values of the Variables. The magnitude of a Variable only effects the type carried by its shuttle, and in no way influences the movements of the shuttle as a whole

The machine, guided by the formula-paper, now causes the shuttle-boxes to rotate until the first shuttles of both inner and outer boxes come opposite to a shuttle-race. The two shuttles are then drawn along the race to a position near the index; and certain slides are released, which move forward until stopped by striking the type carried by the outer shuttle. The slides in question will then have moved distances corresponding to the simple index numbers of the corresponding digits of the Variables b. In the particular case under consideration, the first four slides will therefore move 3, 0, 7, and 1 units respectively, the remainder of the slides indicating zero by moving 50 units (see Table 1). Another slide moves *in the opposite direction* until stopped by the first type of the inner shuttle, making a movement proportional to the simple index number of the first digit of the multiplier a—in this case 14. As the index is attached to the last-mentioned slide, and partakes of its motion, the *relative* displacements of the index and each of the four slides are respectively $3 + 14$, $0 + 14$, $7 + 14$, and $1 + 14$ units (that is, 17, 14, 21, and 15 units), so that pointers attached to the four slides, which normally point to zero on the index, will now point respectively to the 17th, 14th, 21st and 15th divisions of the index. Consulting Table 3, we find that these divisions correspond to the partial products 72, 9, 27, and 18. In the index the partial products are expressed mechanically by movable blades placed at the intervals shown in column 2 of the third table. Now, the duration of the first movement of any blade is as the unit figure of the partial product which it represents, so that the movements of the blades concerned in the present case will be as the numbers 2, 9, 7, and 8, which movements are conveyed by the pointers to the mill, causing it to register the number 2978. A carriage near the index now moves one step to effect multiplication by 10, and then the blades partake of a second movement, this time transferring the tens' figures of the partial products (i.e. 7, 0, 2, and 1) to the mill, which completes the addition of the units' and tens' figures thus:

$$\begin{array}{r} 2978 \\ 7021 \\ \hline 73188 \end{array}$$

the result being the product of the multiplicand b by the first digit of the multiplier a. After this the index makes a rapid reciprocating movement, bringing its slide into contact with the second type of the inner shuttle (which represents the figure 2 in the quantity a), and the process just described is repeated for this and the subsequent figures of the multiplier a until the whole product ab is found. The shuttles are afterwards replaced in the shuttle-boxes, the latter being then rotated until the second shuttles of both boxes are opposite to the shuttle-race. These shuttles are brought to the index, as in the former case, and the product of their Variables (21893×823) is obtained, which, being added to the previous product (that product having been purposely retained in the mill), gives the required value of x. It may be mentioned that the position of the decimal point in a product is determined by special mechanism which is independent of both mill and index.

Most of the movements mentioned above, as well as many others, are derived from a set of cams placed on a common shaft parallel to the driving-shaft; and all movements so derived are under the control of the formula-paper.

The ordinals in Table 1 are not mathematically important, but refer to special mechanism which cannot be described in this paper, and are included in the tables merely to render them complete.

The sum of two products is obtained by retaining the first product in the mill until the second product is found—the mill will then indicate their sum. By reversing the direction of rotation of the mill before the second product is obtained, the difference of the products results. Consequently, by making the multiplier unity in each case, simple addition and subtraction may be performed.

In designing a calculating machine it is a matter of peculiar difficulty and of great importance to provide for the expeditious carrying of tens. In most machines the carryings are performed in rapid succession; but Babbage invented an apparatus (of which I have been unable to ascertain the details) by means of which the machine could 'foresee' the carryings and act on the foresight. After several years' work on the problem, I have devised a method in which the carrying is practically in complete mechanical independence of the adding process, so that the two movements proceed simultaneously. By my method the sum of m numbers of n figures would take $9m + n$ units of time. In finding the product of two numbers of twenty figures each, forty additions are required (the units' and tens' figures of the partial products being added separately). Substituting the values 40 and 20 for m and n, we get $9 \times 40 + 20 = 380$, or $9\frac{1}{2}$ time-units for each addition—the time-unit being the period required to move a figure-wheel through $\frac{1}{10}$ revolution. With Variables of 20 figures each the quantity n has a constant value of 20, which is the number of units of time required by the machine to execute any carrying which has not been performed at the conclusion of an indefinite number of additions. Now, if the carryings were performed in succession, the time required could not be less than $9 + n$, or 29 units for each addition, and is, in practice, considerably greater.*

In ordinary calculating machines division is accomplished by repeated subtractions of the divisor from the dividend. The divisor is subtracted from the figures of the dividend representing the higher powers of ten until the remainder is less than the divisor. The divisor is then moved one place to the right, and the subtraction proceeds as before. The number of subtractions performed in each case denotes the corresponding figure of the quotient. This is a very simple and convenient method for ordinary calculating machines; but it scarcely meets the requirements of an Analytical Machine. At the same time, it must be observed that Babbage used this method, but found it gave rise to many mechanical complications.

My method of dividing is based on quite different principles, and to explain it I must assume that the machine can multiply, add, or subtract any of its Variables; or, in other words, that a formula-paper can be prepared which could direct the machine to evaluate any specified function (which does not contain the sign of division or its equivalent) for given values of its variables.

Suppose, then, we wish to find the value of p/q for particular values of p and q which have been communicated to the machine. Let the first three figures of q be represented by f, and let A be the reciprocal of f, where A is expressed as a decimal of 20 figures. Multiplying the numerator and denominator of the fraction by A, we have $(Ap)/(Aq)$, where Aq must give a number of the form 100 . . . because $Aq = q/f$. On placing the decimal point after the unit, we have unity plus a small decimal. Represent this decimal by x: then

$$\frac{p}{q} = \frac{Ap}{1 + x} \text{ or } Ap\,(1 + x)^{-1}$$

Expanding by the binomial theorem

*For further notes on the problem of the carrying of tens, see C. Babbage: 'Passages from the Life of a Philosopher', p. 114, etc.

Brian Randell, The Computer Journal, Volume 14, Issue 3, pp. 317–326, 1971. © 1971, British Computer Society, reproduced with permission

Ludgate's Analytical Machine of 1909

Brian Randell, The Computer Journal, Volume 14, Issue 3, pp. 317–326, 1971. © 1971, British Computer Society, reproduced with permission

Ludgate's Analytical Machine of 1909

(1) $\dfrac{p}{q} = Ap(1 - x + x^2 - x^3 + x^4 - x^5 + \text{etc.})$,

or

(2) $\dfrac{p}{q} = Ap(1 - x)(1 + x^2)(1 + x^4)(1 + x^8)$, etc.

The series (1) converges rapidly, and by finding the sum as far as x^{10} we obtain the correct result to at least twenty figures; whilst the expression (2) gives the result correctly to at least thirty figures. The position of the decimal point in the quotient is determined independently of these formulae. As the quantity A must be the reciprocal of one of the numbers 100 to 999, it has 900 possible values. The machine must, therefore, have the power of selecting the proper value for the quantity A, and of applying that value in accordance with the formula. For this purpose the 900 values of A are stored in a cylinder—the individual figures being indicated by holes of from one to nine units deep at its periphery. When division is to be performed, this cylinder is rotated, by a simple device, until the number A (represented on the cylinder by a row of holes), which is the reciprocal of the first three figures of the divisor, comes opposite to a set of rods. These rods then transfer that number to the proper shuttle, whence it becomes an ordinary Variable, and is used in accordance with the formula. It is not necessary that every time the process of division is required the dividing formula should be worked out in detail in the formula-paper. To obviate the necessity of so doing the machine is provided with a special permanent *dividing cylinder*, on which this formula is represented in the proper notation of perforations. When the arrangement of perforations on the formula-paper indicates that division is to be performed, and the Variables which are to constitute divisor and dividend, the formula-paper then allows the dividing cylinder to usurp its functions until that cylinder has caused the machine to complete the division.

It will be observed that, in order to carry out the process of division, the machine is provided with a small table of numbers (the numbers A) which it is able to consult and apply in the proper way. I have extended this system to the logarithmic series, in order to give to that series a considerable convergency; and I have also introduced a *logarithmic cylinder* which has the power of working out the logarithmic formula, just as the dividing cylinder directs the dividing process. This system of auxiliary cylinders and tables for special formulae may be indefinitely extended.

The machine prints all results, and, if required, the data, and any noteworthy values which may transpire during the calculation. It may be mentioned, too, that the machine may be caused to calculate and print, quite automatically, a table of values—such, for instance, as a table of logs, sines, squares, etc. It has also the power of recording its results by a system of perforations on a sheet of paper, so that when such a *number-paper* (as it may be called) is replaced in the machine, the latter can 'read' the numbers indicated thereon, and inscribe them in the shuttles reserved for the purpose.

Among other powers with which the machine is endowed is that of changing from one formula to another as desired, or in accordance with a given mathematical law. It follows that the machine need never be idle; for it can be set to tabulate successive values of any function, while the work of the tabulation can be suspended at any time to allow of the determination by it of one or more results of greater importance or urgency. It can also 'feel' for particular events in the progress of its work—such, for instance, as a change of sign in the value of a function, or its approach to zero or infinity; and it can make any pre-arranged change in its procedure, when any such event occurs. Babbage dwells on these and similar points, and explains their bearing on the automatic solution (by approximation) of an equation of the nth degree;* but I have

not been able to ascertain whether his way of attaining these results has or has not any resemblance to my method of so doing.

The Analytical Machine is under the control of two key-boards, and in this respect differs from Babbage's Engine. The upper key-board has ten keys (numbered 0 to 9), and is a means by which numbers are communicated to the machine. It can, therefore, undertake the work of the number-paper previously mentioned. The lower key-board can be used to control the working of the machine, in which case it performs the work of a formula-paper. The key-boards are intended for use when the nature of the calculation does not warrant the preparation of a formula-paper or a number-paper, or when their use is not convenient. An interesting illustration of the use of the lower key-board is furnished by a case in which a person is desirous of solving a number of triangles (say) of which he knows the dimensions of the sides, but has not the requisite formula-paper for the purpose. His best plan is to put a plain sheet of paper in the controlling apparatus, and on communicating to the machine the known dimensions of one of the triangles by means of the upper key-board, to guide the machine by means of the lower key-board to solve the triangle in accordance with the usual rule. The manipulations of the lower key-board will be recorded on the paper, which can then be used as a formula-paper to cause the machine automatically to solve the remaining triangles. He can communicate to the machine the dimensions of these triangles individually by means of the upper key-board; or he may, if he prefers so doing, tabulate the dimensions in a number-paper, from which the machine will read them of its own accord. The machine is, therefore, able to 'remember', as it were, a mathematical rule; and having once been shown how to perform a certain calculation, it can perform any similar calculation automatically so long as the same paper remains in the machine.

It must be clearly understood that the machine is designed to be quite automatic in its action, so that a person almost entirely ignorant of mathematics could use it, in some respects, as successfully as the ablest mathematician. Suppose such a person desired to calculate the cosine of an angle, he obtains the correct result by inserting the formula-paper bearing the correct label, depressing the proper number-keys in succession to indicate the magnitude of the angle, and starting the machine, though he may be quite unaware of the definition, nature, or properties of a cosine.

While the machine is in use its central shaft must be maintained at an approximately uniform rate of rotation—a small motor might be used for this purpose. It is calculated that a velocity of three revolutions per second would be safe; and such a velocity would ensure the multiplication of any two Variables of twenty figures each in about 10 seconds, and their addition or subtraction in about three seconds. The time taken to divide one Variable by another depends on the degree of convergency of the series derived from the divisor, but $1\frac{1}{2}$ minutes may be taken as the probable maximum. When constructing a formula-paper, due regard should therefore be had to the relatively long time required to accomplish the routine of division; and it will, no doubt, be found advisable to use this process as sparingly as possible. The determination of the logarithm of any number would take two minutes, while the evaluation of a^n (for any value of n) by the expotential theorem, should not require more than $1\frac{1}{2}$ minutes longer—all results being of twenty figures.†

The machine, as at present designed, would be about 26 inches long, 24 inches broad, and 20 inches high; and it would therefore be of a portable size. Of the exact dimensions of Babbage's

---

*C. Babbage: 'Passages from the Life of a Philosopher', p. 131.
†The times given include that required for the selection of the Variables to be operated on.

Engine I have no information; but evidently it was to have been a ponderous piece of machinery, measuring many feet in each direction. The relatively large size of this engine is doubtless due partly to its being designed to accommodate the large number of one thousand Variables of fifty figures each, but more especially to the fact that the Variables were to have been stored on columns of wheels, which, while of considerable bulk in themselves, necessitated somewhat intricate gearing arrangements to control their movements. Again, Babbage's method of multiplying by repeated additions, and of dividing by repeated subtractions, though from a mathematical point of view very simple, gave rise to very many mechanical complications.*

To explain the power and scope of an Analytical Machine or Engine, I cannot do better than quote the words of the Countess of Lovelace: 'There is no finite line of demarcation which limits the powers of the Analytical Engine. These powers are coextensive with the knowledge of the laws of analysis itself, and need be bounded only by our acquaintance with the latter. Indeed, we may consider the engine as the material and mechanical representative of analysis, and that our actual working powers in this department of human study will be enabled more effectually than heretofore to keep pace with our theoretical knowledge of its principles and laws, through the complete control which the engine gives us over the executive manipulations of algebraical and numerical symbols.'†

A Committee of the British Association which was appointed to report on Babbage's Engine stated that, 'apart from the question of its saving labour in operations now possible, we think the existence of such an instrument would place within reach much which, if not actually impossible, has been too close to the limits of human skill and endurance to be practically available'.‡

In conclusion, I would observe that of the very numerous branches of pure and applied science which are dependent for their development, record, or application on the dominant science of mathematics, there is not one of which the progress would not be accelerated, and the pursuit would not be facilitated, by the complete command over the numerical interpretation of abstract mathematical expressions, and the relief from the time-consuming drudgery of computation, which the scientist would secure through the existence of machinery capable of performing the most tedious and complex calculations with expedition, automatism, and precision.

Appendix 2

(Reprinted from *Nature* 81, 2070 (1 July 1909) pp. 14-15)

A NEW ANALYTICAL ENGINE

The April number of the Scientific Proceedings of the Royal Dublin Society contains an interesting and very original paper by Mr. Percy E. Ludgate on a proposed analytical machine. Of all calculating machines, the analytical machine or engine is the most comprehensive in its powers. Cash till reckoners and adding machines merely add or add and print results. Arithmometers are used for multiplying and dividing, which they really only accomplish by rapidly repeated addition or subtraction, with the exception alone, perhaps, of the arithmometer of Bollée, which, in a way, works by means of a mechanical multiplication table. Difference engines originated by Babbage produce and print tables of figures of almost any variety, but the process is one of addition of successive differences. The analytical engine proposed by Babbage was intended

*See Report Brit. Assoc., 1878, p. 100.
†R. Taylor's 'Scientific Memoirs', 1843, vol. iii., p. 696.
‡Report Brit. Assoc., 1878, p. 101.

to have powers of calculation so extensive as to seem a long way outside the capacity of mere mechanism, but this was to be brought about by the use of operation cards supplied by the director or user, which, like the cards determining the pattern in a Jacquard loom, should direct the successive operations of the machine, much as the timing cam of an automatic lathe directs the successive movements of the different tools and feeding and chucking devices. However elaborate the mechanism of Babbage, if completed, might have been, the individual elements of operation would, so far as the writer has been able to understand it, have been actually operations of addition or subtraction only, and, with the exception of the method of multiplication created by Bollée, the writer does not recall any case in which mechanism has been used to compute numerical results except by the use of the processes of addition or subtraction, simple or cumulative. Of course, harmonic analysers and other instruments depending on geometry are not included in the category of machines which operate on numbers.

The simplicity of the logarithmic method of multiplying must have made many inventors regret the inherent incommensurability of the function to any simple base, or, if commensurability is attained for any particular number and its powers by the use of an incommensurable base, the incommensurability of the corresponding logarithms of numbers prime to those first selected. On this account the writer has always imagined that the logarithmic method was unsuited to mechanism, or, if applied at all, could only be so applied at the expense of complication, which would more than compensate for the directness of the process of logarithmic multiplication.

Mr. Ludgate, however, in effect, uses for each of the prime numbers below ten a logarithmic system with a different incommensurable base, which as a fact never appears, and is able to take advantage of the additive principle, or, rather, it is so applied that the machine may use it. These mixed or Irish logarithms, or index numbers, as the author calls them, are very surprising at first, but, if the index numbers of zero be excepted, it is not difficult to follow the mode by which they have been selected. The index numbers of the ten digits are as follows:

Digit	0	1	2	3	4	5	6	7	8	9
Index number	50	0	1	7	2	23	8	33	3	14

When two numbers are to be multiplied, the index numbers of the several digits are mechanically added to the index numbers of each of the digits of the other, and, the process of carrying the tens being carried on simultaneously, the time required is very small. For instance, the author gives as an example the multiplication of two numbers of 20 digits each, which will require 40 of these additions, which he shows will require $9\frac{1}{2}$ time units if a time unit is one-tenth of the time of revolution of a figure wheel.

Unfortunately, while the principle on which the proposed machine is to work is described, only the barest idea of the mechanical construction is given, so that it is difficult to judge of the practicability of the intended construction. Whatever this may be, the originality of the method of mixed commensurable logarithms to incommensurable bases seems to the writer so great and the conception so bold as to be worthy of special attention.

Division has hitherto always been effected by the process of rapid but repeated subtraction, following in this respect the method practised with pencil and paper. Having discovered how to harness the logarithm to mechanism, Mr. Ludgate would, it would be expected, have managed to effect division by a logarithmic method, and possibly he could have done so, but here again he has left the beaten track, and by his ingenuity has made division a direct, and not, as hitherto, an indirect or trial-and-error process. Starting with a table of reciprocals of all numbers from 100 to 999, which in a mechanical form is

Ludgate's Analytical Machine of 1909

intended to be stored in the machine, he imagines both numerator and denominator of the required fraction p/q to be multiplied by the reciprocal A of the first three digits of q so as to become $(Ap)/(Aq)$. Aq must, then, in every case begin with the digits 100, and it may be written $1 + x$, where x is a small fraction. Then

$$\frac{p}{q} = Ap(1 - x)(1 + x^2)(1 + x^6)(1 + x^8)\ldots$$

a highly convergent series of which five terms will give a result correct to twenty figures at least, and so division is intended to be effected by a process of direct multiplication.

Until more detail as to the proposed construction and drawings are available it is not possible to form any opinion as to the practicability or utility of the machine as a whole, but it is to be hoped that if the author receives, as he deserves,

encouragement to proceed with his task, he will not allow himself to become swamped in the complexity which must be necessary if he aims at the wide generality of a complete analytical engine. If he will, in the first instance, produce his design for a machine of restricted capacity, even if it does no more than an arithmometer, he will, by demonstrating its practicability and advantages be more likely to be enabled to proceed step by step to the more perfect instrument than he will if, as Babbage did, he imagines his whole machine at once. In the writer's opinion, the ingenuity required to arrange a complete analytical engine is really in great part misplaced. Such a machine can only be used and kept in order by someone who really understands it, and it would seem to the writer of this notice more practicable to allow the user's attention to replace the action of operation cards, and leave to the machine the more direct numerical evaluations.

C. V. Boys

References

AIKEN, H. H. (1937?). Proposed Automatic Calculating Machine. (Previously unpublished memorandum, undated, but bearing an unknown recipient's hand-written notation 'Prospectus of Howard Aiken, November 4, 1937'. Printed in *IEEE Spectrum*, August 1946, pp. 62-69.)

BABBAGE, C. (1851). Calculating Engines. Chapter 13 of *The Exposition of 1851*, London: J. Murray (Reprinted in Morrison and Morrison, 1961).

BABBAGE, H. P. (1888). On the Mechanical Arrangements of the Analytical Engine of the late Charles Babbage, *Report of the Brit. Assoc. for the Advancement of Science*, pp. 616-617.

BAXANDALL, D. (1926). Calculating Machines and Instruments, *Catalogue of the Collections in the Science Museum*, London: His Majesty's Stationery Office.

CHASE, G. C. (1952). History of Mechanical Computing Machinery, *Proceedings of the A.C.M. National Meeting*, Pittsburgh 2-3 May 1952.

HOFFMAN, W. (1962). Entwicklungsbericht und Literaturzusammenstellung über Ziffern-Rechenautomaten, *Digitale Informationswandler*, Braunschweig: Vieweg, pp. 650-717.

LUDGATE, P. E. (1914). Automatic Calculating Engines, *Napier Tercentenary Celebration: Handbook of the Exhibition*, Edinburgh: Royal Society of Edinburgh.

MENABREA, L. F. (1843). Sketch of the Analytical Engine Invented by Charles Babbage. With notes upon the Memoir by the Translator, Ada Augusta, Countess of Lovelace, *Taylor's Scientific Memoirs*, Vol. 3. (Reprinted in Morrison and Morrison, 1961.)

MERRIFIELD, C. W. (1878). Report of the Committee appointed to consider the expense of constructing Mr. Babbage's Analytical Machine, and of printing Tables by its means, *Report of the Brit. Assoc. for the Advancement of Science*, pp. 92-102.

MORRISON, P., and MORRISON, E. (1961). *Charles Babbage and his Calculating Engines, Selected Writings by Charles Babbage and Others*, New York: Dover Publications.

WILKES, M. V. (1956). *Automatic Digital Calculators*, London: Methuen.

1

Speculations on Percy Ludgate's Difference Engine

Brian Coghlan

Abstract In Percy Ludgate's 1914 paper for the Napier Tercentenary Exhibition he briefly mentioned that he had designed a simplified difference engine. This paper speculates on how that might have worked. This paper is a work in progress.

Index Terms—**Percy Ludgate, Analytical Machine, Difference Engine**

I. INTRODUCTION

P ERCY Edwin Ludgate (1883-1922) is notable as the second person to publish a design for an Analytical Machine [1][2], the first after Charles Babbage's "Analytical Engine" [4]. An analytical machine is equivalent to a general-purpose computer, and in theory can be programmed to solve any solvable problem. It is "Turing complete", a term invented to reflect Alan Turing's contribution to the theory of computing. There were only two mechanical designs before the electronic computer era: in 1843 Babbage's "analytical engine", then in 1909 Ludgate's very different "analytical machine".

Babbage only started work on his Analytical Engine in c.1834 when his intensive efforts to design and make a difference engine faltered. This machine was intended to evaluate polynomials using Newton's method of divided differences in order to produce mathematical and nautical tables. In 1914 Ludgate published a paper for the Napier Tercentenary Exhibition [3] (hereafter called "Ludgate 1914") in which he briefly mentioned that he had designed a simplified difference engine. This paper speculates on how that might have worked.

II. DIFFERENCE ENGINES

II.1. Babbage's Difference Engine

Charles Babbage is shown in Figure 2(a), drawn late in his life. He is historically important as the first person to design a difference engine and then also the first to design an analytical engine [22]. From Ludgate 1914:

> The first great automatic calculating machine was invented by Charles Babbage. He called it a "difference-engine," and commenced to construct it about the year 1822. The work was continued during the following twenty years, the Government contributing about £17,000 to defray its cost, and Babbage himself a further sum of about £6000. At the end of that time the construction of the engine, though nearly finished, was unfortunately abandoned owing to some misunderstanding with the Government. A portion of this engine is exhibited in South Kensington Museum, along with other examples of Babbage's work. If the engine had been finished, it would have contained seven columns of wheels, with twenty wheels in each column (for computing with six orders of differences), and also, a contrivance for stereotyping the tables calculated by it. A machine of this kind will calculate a sequence of tabular numbers automatically when its figure-wheels are first set to correct initial values.

Babbage actually designed two difference engines. His difference engine no.1 operated on 20-digit numbers and sixth-order differences. His difference engine no.2 operated on 31-digit numbers and seventh-order differences.

To support $(N-1)$-order differences, his difference engines consisted of N columns, each storing one decimal number, where each column J was only able to add the value of column $(J+1)$ to the contents of column J. Column N could only hold a constant, and column 1 would contain the current result of the calculation. The initial condition was that column 1 was set to the initial value X of the polynomial, column 2 to a value derived from the first and higher derivatives of the polynomial at the same value of X, and columns 3 to N to values derived from the first and higher derivatives of the polynomial.

Babbage used the example $f(X) = X^2 + X + 41$ when explaining these principles, illustrating it using Table 1. The 2nd difference (analogous to 2nd derivative) is constant for a quadratic function. The rest of the table can be calculated using only addition. For example, in the last steps, 2 is added to 10 to give 12 (the 1st difference), then 12 is added to 71 to give the final result of $f(X) = 83$. The previous steps are similarly calculated.

Dr.Brian Coghlan is with the School of Computer Science and Statistics, Trinity College Dublin, The University of Dublin, Ireland (e-mail: coghlan@cs.tcd.ie).

BrianCoghlan-Speculations-on-Percy-Ludgates-Difference-Engine-20210918-1550.doc

X	f(X)	1st	2nd
0	41		
		2	
1	43		2
		2	
2	47		2
		6	
3	53		2
		8	
4	61		2
		10	
5	71		2
		12	
6	83		

Table 1 *Charles Babbage's example difference engine calculation*

Babbage's difference engines were entirely mechanical, but never completed. The later reconstruction of difference engine no.2 at the London Science Museum under the guidance of Doron Swade [15] (see Figure 1) demonstrated that it was able to be constructed with the technology of the 1850s. Smaller scale demonstration models have been constructed in Lego [19] and Meccano [20], as well as educational software models, e.g [21] and [24].

Figure 1 *Charles Babbage's Difference Engine No.2 in the Science Museum, London Completed in 2002, with 8,000 parts, 11 feet long, and weighing five tons,*

3

Figure 2 *(a) Charles Babbage, from his Obituary, 4th November 1871, The Illustrated London News*
(b) Percy Ludgate, image reproduced courtesy Brian Randell

II.2. Ludgate's Difference Engine

Percy Ludgate is shown in Figure 2(b), probably taken in the last five years of his life. He was the second person in history to publish (in 1909) a design for an analytical machine. But later, in 1914, he published a paper in the Handbook of the Napier Tercentenary Exhibition [3] in which he stated that he had designed a simple difference engine. From Ludgate 1914:

> The most pleasing characteristic of a difference-engine made on Babbage's principle is the simplicity of its action, the differences being added together in unvarying sequence; but notwithstanding its simple action, its structure is complicated by a large amount of adding mechanism—a complete set of adding wheels with carrying gear being required for the tabular number, and every order of difference except the highest order. On the other hand, while the best feature of the analytical engine or machine is the Jacquard apparatus (which, without being itself complicated, may be made a powerful instrument for interpreting mathematical formulae), its weakness lies in the diversity of movements the Jacquard apparatus must control. Impressed by these facts, and with the desirability of reducing the expense of construction, I designed a second machine in which are combined the best principles of both the analytical and difference types, and from which are excluded their more expensive characteristics. By using a Jacquard I found it possible to eliminate the redundancy of parts hitherto found in difference-engines, while retaining the native symmetry of structure and harmony of action of machines of that class. My second machine, of which the design is on the point of completion, will contain but one set of adding wheels, and its movements will have a rhythm resembling that of the Jacquard loom itself. It is primarily intended to be used as a difference-machine, the number of orders of differences being sixteen. Moreover, the machine will also have the power of automatically evaluating a wide range of miscellaneous formulae.

Ludgate 1914 only mentions the "Jacquard apparatus" (sequencer) and "one set of adding wheels" (accumulator). There is no mention of the novel storage or multiplier mechanisms of his analytical machine, so one might reasonably presume they were absent. The implication is that Ludgate traded physical complexity for programming, another novel Ludgate idea, and one that predated reduced instruction set computers (RISC) by eighty years.

It is clear the engine adopted an accumulator architecture. One might reasonably presume his accumulator could do subtraction by reverse rotation of the adding wheels, as did the "Mill" of his analytical machine. But there is no mention of any multiplication and division.

Jacquard looms could not do conditional operations, and neither could Babbage's difference engines (difference engines don't need looping constructs). Ludgate 1914 doesn't preclude his design being in essence an analytical engine behaving like a difference engine, but neither does it encourage it. The simpler Ludgate made the engine, the faster it could cycle (a la RISC machines), trading physical complexity not only for programming but also for speed. If there was no conditionality (no testing of conditions), and only ripple carry (2 x slower ripple carry but maybe 2 x faster cycling), it might avoid the penalties associated with ripple carry.

Babbage's difference engine no.2 was hand-cranked, with four turns of the crank per machine cycle. Each cycle consisted of a sequence of 4 simultaneous 31-digit additions, 4 ripple carry corrections, 3 simultaneous 31-digit additions, and 3 ripple carry corrections, i.e. 7 additions with carries excluding print and typeset operations. It takes about 8 seconds per cycle [15]. By way of

Speculations on Percy Ludgate's Difference Engine

4

comparison, Ludgate's analytical engine appears to have had a cycle time of $\frac{1}{3}$ of a second, with each Mill wheel counting as fast as one step per $\frac{1}{100}$ of a second. If Ludgate's difference engine only performed addition and subtraction, then its accumulator wheels could count at this rate, allowing an addition time of about $\frac{1}{10}$ of a second without ripple carry, or perhaps $\frac{1}{5}$ of a second with ripple carry. Therefore the 7 additions with carries would take about 1.2 seconds, faster but not very much faster than Babbage's, but with a much simpler mechanism, and the ability to operate automatically from programs on Jacquard cards.

One big puzzle is whether there was proper storage. Ludgate stated there would be 16 orders of difference, so at minimum 16 temporary registers. One possibility would be a set of 16 registers constantly moving like a conveyor belt. Perhaps Ludgate poached from his analytical machine ideas of shuttles in storage cylinders, perhaps with shuttles in a constantly rotating ring, or with shuttles in a constantly rotating storage cylinder (with no address selection), somewhat reminiscent of Babbage's 1932 introduction of a "feedback mechanism" to his Difference Engine [25], the idea that ultimately appears to have spawned his Analytical Engine.

The kernel of the operations could be as in Algorithm 1.

```
repeat
    acc = initial_value
    for d in row_of_differences
        acc = acc + d
        d = acc
    print acc
```

Algorithm 1 *Speculative example operations of Ludgate's difference engine calculation*

So to calculate the squares one might initialize with row of differences = [2, -1, 0], yielding a printout of 1, 4, 9, 16. The temporary storage for `row_of_differences` would not do any calculation, but it would need to present its contents to the accumulator in such a way that the contents were added to it, and then be set to the value of the accumulator.

With 16 orders of difference it might be useful to scale (shift towards most or least significant digit) the accumulator at each iteration [23], and/or to employ fixed-point arithmetic. Doron Swade [18] has said "The position of the decimal in the Difference Engines is notional, i.e. there is nothing mechanical that marks its location. In the case of DE2, which has vertical columns for multidigit numbers, the position of the decimal point is along a notional fixed horizontal line across the columns." This approach is equally applicable to Ludgate's difference engine.

With no ordinals nor logarithmic indexes, temporary storage shuttles could hold decimals on rods, avoiding much of the complexity of Ludgate's analytical machine. Furthermore, if the temporary storage held quantities 0..10 for each difference rather than 0..9, then additions could give 20 with just a single carry, rather than two rounds of carry. This would avoid ripple carry until the very end, and virtually double the speed.

Another puzzle is what exactly the program steps might execute. An initial instruction might carry the initial value of $f(X)$ and the differences (with at minimum just the 1st difference and the rest zero), perhaps with these operands in succeeding rows of the Jacquard cards, until a hole marked the last operand. Subsequent instructions might perform the requisite additions or subtractions until the calculation was complete, followed by printing of the result. À la unrolled iterations, further in-line instructions could repeat this pattern of arithmetic and printing instructions until a whole table had been calculated and printed. The nett outcome would be a programmable difference engine. The Jacquard cards would be consumed in one direction with no looping, just as in Jacquard looms. From Ludgate 1914: "its movements will have a rhythm resembling that of the Jacquard loom itself", which conjures up an image of the engine merrily gobbling up the Jacquard cards.

Ludgate's final two sentences are intriguing:

It is primarily intended to be used as a difference-machine, the number of orders of differences being sixteen. Moreover, the machine will also have the power of automatically evaluating a wide range of miscellaneous formulae.

Randell has highlighted that the words "It is primarily intended" indicate that Ludgate had a wider view of its powers, that it was not strictly a difference engine, that it could do more than a pure difference engine could. But Ludgate's wording does not hint at any addressable storage, nor conditional branching. If they were absent, then it was just a special-purpose machine (and faster without conditional branching), a sort of "formulae evaluation engine", certainly not an analytical engine. An analytical engine can be programmed to execute any of its instruction set including conditional instructions in any order (not a fixed set of operations in a fixed order), and has addressable storage with programmable load/store. A pure difference engine cannot do any of those operations, but perhaps the machine mentioned in Ludgate 1914 could do some subset of those operations. Lack of hints does not mean absence, we cannot assume, we can only speculate

5

III. Concluding Remarks

The above must not be construed as anything but speculations. The paucity of information in Ludgate 1914 precludes any other complexion. What Ludgate states does, however, imply an entirely novel concept at the time.

IV. Acknowledgements

My thanks to David McQuillan [17] for a number of the above speculations, and Brian Randell [6][7] who both kindled discussion of Ludgate's difference engine and contributed fresh insights, and Doron Swade [18] for clarifications.

References,

[1] Percy E.Ludgate, On a Proposed Analytical Machine, Scientific Proceedings of the Royal Dublin Society, Vol.12, No.9, pp.77–91, 28-Apr-1909. Also reproduced in [6].
[2] C.V.Boys, A new analytical engine, Nature, Vol.81, pp.14-15, Jul-1909. Also reproduced in [6].
[3] Percy E.Ludgate, Automatic Calculating Machines, In: Handbook of the Napier Tercentenary Celebration or modern instruments and methods of calculation, Ed: E.M.Horsburgh, 1914.
[4] Lovelace, A., Sketch of the Analytical Engine invented by Charles Babbage, Esq. By L.F. Menabrea, of Turin, Officer of the Military Engineers, with notes by the Translator, Scientific Memoirs, Ed.R.Taylor, Vol.3, pp.666–731, 1843. Available at:
https://www.scss.tcd.ie/SCSSTreasuresCatalog/literature/TCD-SCSS-V.20121208.870/TCD-SCSS-V.20121208.870.pdf
[5] Baxandall, D., Calculating machines and instruments: Catalogue of the Collection in the Science Museum, Science Museum, London, 1926.
[6] Brian Randell, Ludgate's analytical machine of 1909, The Computer Journal, Vol.14, No.3, pp.317-326, 1971. Available at [14] in file:
"Randell-Ludgates-Analytical-Machine-of-1909-TheComputerJournal-1971-317-326.pdf".
[7] Brian Randell, From analytical engine to electronic digital computer: The contributions of Ludgate, Torres, and Bush, Annals of the History of Computing, Vol.4, No.4, IEEE, Oct. 1982. Available at [14] in file:
"Randell-Contributions-of-Ludgate-Torres-and-Bush-IEEEAnnals-1982-4-4.pdf".
[8] Trinity College Dublin, The John Gabriel Byrne Computer Science Collection. Available at: https://www.scss.tcd.ie/SCSSTreasuresCatalog/
[9] Trinity College Dublin, Percy E. Ludgate Prize in Computer Science. Available at:
https://www.scss.tcd.ie/SCSSTreasuresCatalog/miscellany/TCD-SCSS-X.20121208.002/TCD-SCSS-X.20121208.002.pdf
[10] Brian Coghlan, Percy Ludgate's Analytical Machine, Trinity College Dublin. Available at [14] in file:
"BrianCoghlan-PercyLudgatesAnalyticalMachine-20200617-1330.pdf".
Also updated version. Available at [14] in file: "BrianCoghlan-PercyLudgatesAnalyticalMachine.pdf",
[11] Joseph Marie Jacquard, Jacquard Loom, c.1804. Available at: https://en.wikipedia.org/wiki/Joseph_Marie_Jacquard
[12] Trinity College Dublin, Percy E. Ludgate: Part A1, Prize in Computer Science. Available at [14] in file:
"TCD-SCSS-X.20121208.002-20191027-1533-Draft-partA1.pdf".
[13] Trinity College Dublin, Percy E. Ludgate: Part A2, Extended Discussion. Available at [14] in file:
"TCD-SCSS-X.20121208.002-20191027-1533-Draft-partA2.pdf".
[14] Trinity College Dublin, Percy E. Ludgate folder. Available at:
https://www.scss.tcd.ie/SCSSTreasuresCatalog/miscellany/TCD-SCSS-X.20121208.002/
[15] Doron Swade, The Construction of Charles Babbage's Difference Engine No. 2, Vol.27, No.3, IEEE Annals of the History of Computing, July 2005. Available at: https://www.quora.com/How-fast-in-terms-of-%E2%80%98flops-per-second%E2%80%99-was-Babbage%E2%80%99s-analytical-engine
[16] Quora, How fast, in terms of 'flops per second', was Babbage's analytical engine?. Available at:
https://www.quora.com/How-fast-in-terms-of-%E2%80%98flops-per-second%E2%80%99-was-Babbage%E2%80%99s-analytical-engine
[17] David McQuillan, The Feasibility of Ludgate's Analytical Machine. Available at:http://www.fano.co.uk/ludgate/Ludgate.html
[18] Doron Swade, private email communication, 18-Jul-2020.
[19] Andrew Carol, Building A Calculating Machine Using LEGO pieces. Available at:
https://www.cs.princeton.edu/~chazelle/courses/BIB/BabbageEngine.html
[20] Tim Robinson, Robinson's Difference Engine #1. Available at: https://www.meccano.us/difference_engines/rde_1/
[21] University of Warwick, Modelling Babbage's Difference Engine, Department of Computer Science, . Available at:
https://warwick.ac.uk/fac/sci/dcs/research/em/publications/web-em/02/babbage.pdf
[22] Bruce Collier, Little Engines That Could've: The Calculating Machines of Charles Babbage, Ph.D. Thesis, Harvard University, 1970. Available at:
http://robroy.dyndns.info/collier/
[23] Ed Thelen, Babbage Difference Engine #2: Scaling to permit dealing with fractions. Available at: http://ed-thelen.org/bab/bab-intro.html#Scaling
[24] Ed Thelen, Babbage Difference Engine #2: a JavaScript non-pipelined emulation. Available at: http://ed-thelen.org/bab/bab-diff-JavaScript.html
[25] Charles Babbage, letter to John Herschel, 15-Dec-1832, John Herschel Papers, 2.275, Royal Society of London.

124 SECTION D

II. Automatic Calculating Machines. By P. E. LUDGATE.

AUTOMATIC calculating machines on being actuated, if necessary, by uniform motive power, and supplied with numbers on which to operate, will compute correct results without requiring any further attention. Of course many adding machines, and possibly a few multiplying machines, belong to this category ; but it is not to them, but to machines of far greater power, that this article refers. On the other hand, tide-predicting machines and other instruments that work on geometrical principles will not be considered here, because they do not operate arithmetically. It must be admitted, however, that the true automatic calculating machine belongs to a possible rather than an actual class ; for, though several were designed and a few constructed, the writer is not aware of any machine in use at the present time that can determine numerical values of complicated formulæ without the assistance of an operator.

The first great automatic calculating machine was invented by Charles Babbage. He called it a " difference-engine," and commenced to construct it about the year 1822. The work was continued during the following twenty years, the Government contributing about £17,000 to defray its cost, and Babbage himself a further sum of about £6000. At the end of that time the construction of the engine, though nearly finished, was unfortunately abandoned owing to some misunderstanding with the Government. A portion of this engine is exhibited in South Kensington Museum, along with other examples of Babbage's work. If the engine had been finished, it would have contained seven columns of wheels, with twenty wheels in each column (for computing with six orders of differences), and also a contrivance for stereotyping the tables calculated by it. A machine of this kind will calculate a sequence of tabular numbers automatically when its figure-wheels are first set to correct initial values.

Inspired by Babbage's work, Scheutz of Stockholm made a difference-engine, which was exhibited in England in 1854, and subsequently acquired for Dudley Observatory, Albany, U.S.A. Scheutz's engine had mechanism for calculating with four orders of differences of sixteen figures each, and for stereotyping its results ; but as it was only suitable for calculating tables having small tabular intervals, its utility was limited. A duplicate of this engine was constructed for the Registrar General's Office, London.

In 1848 Babbage commenced the drawings of an improved difference-engine, and though he subsequently completed the drawings, the improved engine was not made.

Babbage began to design his " analytical engine " in 1833, and he put together a small portion of it shortly before his death in 1871. This engine was to be capable of evaluating any algebraic formula, of which a numerical solution is possible, for any given values of the variables. The formula it is desired to evaluate would be communicated to the engine by two sets of perforated cards similar to those used in the Jacquard loom. These cards would cause the engine automatically to operate on the numerical data placed in it, in such a way as to produce the correct result. The mechanism of this

P. E. Ludgate, in: Handbook of the Napier Tercentenary Celebration, E. M. Horsburgh, Ed., Royal Society of Edinburgh, 1914. © 1914, RSE

PORTRAIT OF CHARLES BABBAGE.

[To face p. 124.

engine may be divided into three main sections, designated the "Jacquard apparatus," the "mill," and the "store." Of these the Jacquard apparatus would control the action of both mill and store, and indeed of the whole engine.

The store was to consist of a large number of vertical columns of wheels, every wheel having the nine digits and zero marked on its periphery. These columns of wheels Babbage termed "variables," because the number registered on any column could be varied by rotating the wheels on that column. It is important to notice that the variables could not perform any arithmetical operation, but were merely passive registering contrivances, corresponding to the pen and paper of the human computer. Babbage originally intended the store to have a thousand variables, each consisting of fifty wheels, which would give it capacity for a thousand fifty-figure numbers. He numbered the variables consecutively, and represented them by the symbols V_1, V_2, V_3, V_4 . . . . . V_{1000}. Now, if a number, say 3·14159, were placed on the 10th variable, by turning the wheels until the number appeared in front, reading from top to bottom, we may express the fact by the equation $V_{10} = 3·14159$ or $V_{10} = \pi$. We may equate the symbol of the variable either to the actual number the variable contains, or to the algebraic equivalent of that number. Moreover, in theoretical work it is often convenient to use literal instead of numerical indices for the letters V, and therefore $V_n = ab$ means that the nth variable registers the numerical value of the product of a and b.

The mill was designed for the purpose of executing all four arithmetical operations. If V_n and V_m were any two variables, whose sum, difference, product, or quotient was required, the numbers they represent would first be automatically transferred to the mill, and then submitted to the requisite operation. Finally, the result of the operation would be transferred from mill to store, being there placed on the variable (which we will represent by V_z) destined to receive it. Consequently the four fundamental operations of the machine may be written as follows :—

$$(1) \quad V_n + V_m = V_z.$$
$$(2) \quad V_n - V_m = V_z.$$
$$(3) \quad V_n \times V_m = V_z.$$
$$(4) \quad V_n \div V_m = V_z.$$

Where n, m, and z may be any positive integers, not exceeding the total number of variables, n and m being unequal.

One set of Jacquard cards, called "directive cards," (also called "variable cards") would control the store, and the other set, called "operation cards," would control the mill. The directive cards were to be numbered like the variables, and every variable was to have a supply of cards corresponding to it. These cards were so designed that when one of them entered the engine it would cause the Jacquard apparatus to put the corresponding variable into gear. In like manner every operation card (of which only four kinds were required) would be marked with the sign of the particular operation it could cause the mill to perform. Therefore, if a directive card bearing the number 16 (say) were to enter the engine, it would cause the

P. E. Ludgate, in: Handbook of the Napier Tercentenary Celebration, E. M. Horsburgh, Ed., Royal Society of Edinburgh, 1914. © 1914, RSE

Automatic Calculating Machines

P. E. Ludgate, in: Handbook of the Napier Tercentenary Celebration, E. M. Horsburgh, Ed., Royal Society of Edinburgh, 1914. © 1914, RSE

number on V_{16} to be transferred to the mill or *vice versa* ; and an operation card marked with the sign $\div$ would, on entering the engine, cause the mill to divide one of the numbers transferred to it by the other. It will be observed that the choice of a directive card would be represented in the notation by the substitution of a numerical for a literal index of a V ; or, in other words, the substitution of an integer for one of the indices n, m, and z in the foregoing four examples. Therefore three directive cards strung together would give definite values to n, m, and z, and one operation card would determine the nature of the arithmetical operation, so that four cards in all would suffice to guide the machine to select the two proper variables to be operated on, to subject the numbers they register to the desired operation, and to place the result on a third variable. If the directive cards were numbered 5, 7, and 3, and the operation card marked $+$, the result would be $V_5 + V_7 = V_3$.

As a further illustration, suppose the directive cards are strung together so as to give the following successive values to n, m, and z :—

$$\text{Sequence of values for} \quad n \ . \ . \ . \ 2, 6, 4, 7.$$
$$,, \qquad ,, \qquad m \ . \ . \ . \ 3, 1, 5, 8.$$
$$,, \qquad ,, \qquad z \ . \ . \ . \ 6, 7, 8, 9.$$

Let the sequence of operation cards be

$$+ \ \times \ - \ \div$$

When the cards are placed in the engine, the following results are obtained in succession :—

$$\text{1st operation,} \quad V_2 + V_3 = V_6.$$
$$\text{2nd} \qquad ,, \qquad V_6 \times V_1 = V_7.$$
$$\text{3rd} \qquad ,, \qquad V_4 - V_5 = V_8.$$
$$\text{4th} \qquad ,, \qquad V_7 \div V_8 = V_9.$$

From an inspection of the foregoing it appears that V_1, V_2, V_3, V_4, and V_5 are independent variables, while V_6, V_7, V_8, and V_9 have their values calculated by the engine, and therefore the former set must contain the data of the calculation.

Let $V_1 = a$, $V_2 = b$, $V_3 = c$, $V_4 = d$, and $V_5 = e$, then we have

$$\text{1st operation,} \quad V_2 + V_3 = b + c = V_6.$$
$$\text{2nd} \qquad ,, \qquad V_6 \times V_1 = (b+c)a = V_7.$$
$$\text{3rd} \qquad ,, \qquad V_4 - V_5 = d - e = V_8.$$
$$\text{4th} \qquad ,, \qquad V_7 \div V_8 = \frac{(b+c)a}{d-e} = V_9.$$

Consequently, whatever numerical values of a, b, c, d, and e are placed on variables V_1 to V_5 respectively, the corresponding value of $\frac{a(b+c)}{d-e}$ will be found on V_9, when all the cards have passed through the machine. Moreover, the same set of cards may be used any number of times for different calculations by the same formula.

In the foregoing very simple example the algebraic formula is deduced from a given sequence of cards. It illustrates the converse of the practical procedure, which is to arrange the cards to interpret a given formula, and it also shows that the cards constitute a mathematical notation in themselves.

Seven years after Babbage died a Committee of the British Association appointed to consider the advisability and to estimate the expense of constructing the analytical engine reported that: " We have come to the conclusion that in the present state of the design it is not possible for us to form any reasonable estimate of its cost or its strength and durability." In 1906 Charles Babbage's son, Major-General H. P. Babbage, completed the part of the engine known as the " mill," and a table of twenty-five multiples of π, to twenty-nine figures, was published as a specimen of its work, in the *Monthly Notices of the Royal Astronomical Society*, April 1910.

I have myself designed an analytical machine, on different lines from Babbage's, to work with 192 variables of 20 figures each. A short account of it appeared in the *Scientific Proceedings, Royal Dublin Society*, April 1909. Complete descriptive drawings of the machine exist, as well as a description in manuscript, but I have not been able to take any steps to have the machine constructed.

The most pleasing characteristic of a difference-engine made on Babbage's principle is the simplicity of its action, the differences being added together in unvarying sequence ; but notwithstanding its simple action, its structure is complicated by a large amount of adding mechanism—a complete set of adding wheels with carrying gear being required for the tabular number, and every order of difference except the highest order. On the other hand, while the best feature of the analytical engine or machine is the Jacquard apparatus (which, without being itself complicated, may be made a powerful instrument for interpreting mathematical formulæ), its weakness lies in the diversity of movements the Jacquard apparatus must control. Impressed by these facts, and with the desirability of reducing the expense of construction, I designed a second machine in which are combined the best principles of both the analytical and difference types, and from which are excluded their more expensive characteristics. By using a Jacquard I found it possible to eliminate the redundancy of parts hitherto found in difference-engines, while retaining the native symmetry of structure and harmony of action of machines of that class. My second machine, of which the design is on the point of completion, will contain but *one* set of adding wheels, and its movements will have a rhythm resembling that of the Jacquard loom itself. It is primarily intended to be used as a difference-machine, the number of orders of differences being sixteen. Moreover, the machine will also have the power of automatically evaluating a wide range of miscellaneous formulæ.

P. E. Ludgate. in: Handbook of the Napier Tercentenary Celebration, E. M. Horsburgh, Ed., Royal Society of Edinburgh, 1914. © 1914, RSE

Automatic Calculating Machines

Automatic Calculating Machines

P. E. Ludgate, in: Handbook of the Napier Tercentenary Celebration, E. M. Horsburgh, Ed., Royal Society of Edinburgh, 1914. © 1914, RSE

THE

SCIENTIFIC PROCEEDINGS

OF THE

ROYAL DUBLIN SOCIETY.

Vol. XII. (N.S.), No. 9.　　　　　　　APRIL, 1909.

ON A PROPOSED ANALYTICAL MACHINE.

BY

PERCY E. LUDGATE.

DUBLIN:

PUBLISHED BY THE ROYAL DUBLIN SOCIETY,
LEINSTER HOUSE, DUBLIN.

WILLIAMS AND NORGATE,
14, HENRIETTA STREET, COVENT GARDEN, LONDON, W.C.

1909.

Price Sixpence.

On a proposed Analytical Machine

Percy E. Ludgate, Scientific Proceedings of the Royal Dublin Society, Vol. 12, No. 9, pp. 77–91, 1909. © 1909, RDS

[77]

IX.

ON A PROPOSED ANALYTICAL MACHINE.

By PERCY E. LUDGATE.

(COMMUNICATED BY PROFESSOR A. W. CONWAY, M.A.)

[Read FEBRUARY 23. Ordered for Publication MARCH 9. Published APRIL 28, 1909.]

I PURPOSE to give in this paper a short account of the result of about six years' work, undertaken by me with the object of designing machinery capable of performing calculations, however intricate or laborious, without the immediate guidance of the human intellect.

In the first place I desire to record my indebtedness to Professor C. V. Boys, F.R.S., for the assistance which I owe to his kindness in entering into correspondence with me on the matter to which this paper is devoted.

It would be difficult and very inadvisable to write on the present subject without referring to the remarkable work of Charles Babbage, who, having first invented two Difference Engines, subsequently (about eighty years ago) designed an Analytical Engine, which was shown to be at least a theoretical possibility ; but unfortunately its construction had not proceeded far when its inventor died. Since Babbage's time his Analytical Engine seems to have been almost forgotten; and it is probable that no living person understands the details of its projected mechanism. My own knowledge of Babbage's Engines is slight, and for the most part limited to that of their mathematical principles.

The following definitions of an Analytical Engine, written by Babbage's contemporaries, describe its essential functions as viewed from different standpoints :—

"A machine to give us the same control over the executive which we have hitherto only possessed over the legislative department of mathematics."[1]

"The material expression of any indefinite function of any degree of generality and complexity, such as, for instance :—$F (x, y, z, \log x, \sin y, \&c.)$, which is, it will be observed, a function of all other possible functions of any number of quantities."[2]

---

[1] C. Babbage : " Passages from the Life of a Philosopher," p. 129.
[2] R. Taylor's " Scientific Memoirs," 1843, vol. iii., p. 691.

"An embodying of the science of operations constructed with peculiar reference to abstract number as the subject of those operations."[1]

"A machine for weaving algebraical patterns."[2]

These four statements show clearly that an Analytical Machine "does not occupy common ground with mere 'calculating machines.' It holds a position wholly its own."

In order to prevent misconception, I must state that my work was not based on Babbage's results—indeed, until after the completion of the first design of my machine, I had no knowledge of his prior efforts in the same direction. On the other hand, I have since been greatly assisted in the more advanced stages of the problem by, and have received valuable suggestions from, the writings of that accomplished scholar. There is in some respects a great resemblance between Babbage's Analytical Engine and the machine which I have designed—a resemblance which is not, in my opinion, due wholly to chance, but in a great measure to the nature of the investigations, which tend to lead to those conclusions on which the resemblance depends. This resemblance is almost entirely confined to the more general, abstract, or mathematical side of the question; while the contrast between the proposed structure of the two projected machines could scarcely be more marked.

It is unnecessary for me to prove the possibility of designing a machine capable of automatically solving all problems which can be solved by numbers. The principles on which an Analytical Machine may rest "have been examined, admitted, recorded, and demonstrated."[3] I would refer those who desire information thereon to the Countess of Lovelace's translation of an article on Babbage's Engine, which, together with copious notes by the translator, appears in R. Taylor's "Scientific Memoirs," vol. iii.; to Babbage's own work, "Passages from the Life of a Philosopher"; and to the Report of the British Association for the year 1878, p. 92. These papers furnish a complete demonstration that the whole of the developments and operations of analysis are capable of being executed by machinery.

Notwithstanding the complete and masterly treatment of the question to be found in the papers mentioned, it will be necessary for me briefly to outline the principles on which an Analytical Machine is based, in order that my subsequent remarks may be understood.

An Analytical Machine must have some means of storing the numerical data of the problem to be solved, and the figures produced at each successive

[1] R. Taylor's "Scientific Memoirs," 1843, vol. iii., p. 694.
[2] loc. cit., p. 696.
[3] C. Babbage: "Passages from the Life of a Philosopher," p. 450.

step of the work (together with the proper algebraical signs); and, lastly, a means of recording the result or results. It must be capable of submitting any two of the numbers stored to the arithmetical operation of addition, subtraction, multiplication, or division. It must also be able to select from the numbers it contains the proper numbers to be operated on; to determine the nature of the operation to which they are to be submitted; and to dispose of the result of the operation, so that such result can be recalled by the machine and further operated on, should the terms of the problem require it. The sequence of operations, the numbers (considered as abstract quantities only) submitted to those operations, and the disposition of the result of each operation, depend upon the algebraical statement of the calculation on which the machine is engaged; while the magnitude of the numbers involved in the work varies with the numerical data of that particular case of the general formula which is in process of solution. The question therefore naturally arises as to how a machine can be made to follow a particular law of development as expressed by an algebraic formula. An eminently satisfactory answer to that question (and one utilized by both Babbage and myself) is suggested by the Jacquard loom, in which interesting invention a system of perforated cards is used to direct the movements of the warp and weft threads, so as to produce in the woven material the pattern intended by the designer. It is not difficult to imagine that a similar arrangement of cards could be used in a mathematical machine to direct the weaving of numbers, as it were, into algebraic patterns, in which case the cards in question would constitute a kind of mathematical notation. It must be distinctly understood that, if a set of such cards were once prepared in accordance with a specified formula, it would possess all the generality of algebra, and include an infinite number of particular cases.

I have prepared many drawings of the machine and its parts; but it is not possible in a short paper to go into any detail as to the mechanism by means of which elaborate formulæ can be evaluated, as the subject is necessarily extensive and somewhat complicated; and I must, therefore, confine myself to a superficial description, touching only points of particular interest or importance.

Babbage's Jacquard-system and mine differ considerably; for, while Babbage designed two sets of cards—one set to govern the operations, and the other set to select the numbers to be operated on—I use one sheet or roll of perforated paper (which, in principle, exactly corresponds to a set of Jacquard-cards) to perform both these functions in the order and manner necessary to solve the formula to which the particular paper is assigned. To such a paper I apply the term *formula-paper*. Each row of perforations across the formula-paper directs the machine in some definite step in the process of calculation—such as, for instance, a complete multiplication, including the selection of the numbers to be multiplied together. Of course a single formula-paper can be used for an indefinite number of calculations, provided that they are all of one type or kind (*i.e.* algebraically identical).

In referring to the numbers stored in the machine, the difficulty arises as to whether we refer to them as mere numbers in the restricted arithmetical sense, or as quantities, which, though always expressed in numerals, are capable of practically infinite variation. In the latter case they may be regarded as true mathematical variables. It was Babbage's custom (and one which I shall adopt) when referring to them in this sense to use the term "Variable" (spelt with capital V), while applying the usual meanings to the words "number" and "variable."

In my machine each Variable is stored in a separate shuttle, the individual figures of the Variable being represented by the relative positions of protruding metal rods or "type," which each shuttle carries. There is one of these rods for every figure of the Variable, and one to indicate the sign of the Variable. Each rod protrudes a distance of from 1 to 10 units, according to the figure or sign which it is at the time representing. The shuttles are stored in two co-axial cylindrical shuttle-boxes, which are divided for the purpose into compartments parallel to their axis. The present design of the machine provides for the storage of 192 Variables of twenty figures each; but both the number of Variables and the number of figures in each Variable may, if desired, be greatly increased. It may be observed, too, that the shuttles are quite independent of the machine, so that new shuttles, representing new Variables, can be introduced at any time.

When two Variables are to be multiplied together, the corresponding shuttles are brought to a certain system of slides called the *index*, by means of which the machine computes the product. It is impossible precisely to describe the mechanism of the index without drawings; but it may be compared to a slide-rule on which the usual markings are replaced by movable blades. The index is arranged so as to give several readings simultaneously. The numerical values of the readings are indicated by periodic displacements of the blades mentioned, the duration of which displacements are recorded in units measured by the driving shaft on a train of wheels called the *mill*, which performs the carrying of tens, and indicates the final product. The product can be transferred from thence to any shuttle, or to two shuttles simultaneously, provided that they do not

Percy E. Ludgate, Scientific Proceedings of the Royal Dublin Society, Vol. 12, No. 9, pp. 77–91, 1909. © 1909, RDS

On a proposed Analytical Machine

belong to the same shuttle-box. The act of inscribing a new value in a shuttle automatically cancels any previous value that the shuttle may have contained. The fundamental action of the machine may be said to be the multiplying together of the numbers contained in any two shuttles, and the inscribing of the product in one or two shuttles. It may be mentioned here that the fundamental process of Babbage's Engine was not multiplication but addition.

Though the index is analogous to the slide-rule, it is not divided logarithmically, but in accordance with certain *index numbers*, which, after some difficulty, I have arranged for the purpose. I originally intended to use the logarithmic method, but found that some of the resulting intervals were too large; while the fact that a logarithm of zero does not exist is, for my purpose, an additional disadvantage. The index numbers (which I believe to be the smallest whole numbers that will give the required results) are contained in the following tables:—

[TABLES.

TABLE I.

Unit.	Simple Index No.	Ordinal.
0	50	9
1	0	0
2	1	1
3	7	4
4	2	2
5	23	7
6	8	5
7	33	8
8	3	3
9	14	6

TABLE 2.

Partial product.	Comp. Index No.	Partial product.	Comp. Index No.	Partial product.	Comp. Index No.
1	0	15	30	36	16
2	1	16	4	40	26
3	7	18	15	42	41
4	2	20	25	45	37
5	23	21	40	48	11
6	8	24	10	49	66
7	33	25	46	54	22
8	3	27	21	56	36
9	14	28	35	63	47
10	24	30	31	64	6
12	9	32	5	72	17
14	34	35	56	81	28

Comp. index numbers of zero :—50, 51, 52, 53, 57, 58, 64, 73, 83, 100.

Percy E. Ludgate, Scientific Proceedings of the Royal Dublin Society, Vol. 12, No. 9, pp. 77–91, 1909. © 1909, RDS

On a proposed **Analytical Machine**

TABLE 3.

Comp. Index No.		Partial product.	Comp. Index No.		Partial product.
0	..	1	34	..	14
1	..	2	35	..	28
2	..	4	36	..	56
3	..	8	37	..	45
4	..	16	38	..	—
5	..	32	39	..	—
6	..	64	40	..	21
7	..	3	41	..	42
8	..	6	42	..	—
9	..	12	43	..	—
10	..	24	44	..	—
11	..	48	45	..	—
12	..	—	46	..	25
13	..	—	47	..	63
14	..	9	48	..	—
15	..	18	49	..	—
16	..	36	50	..	0
17	..	72	51	..	0
18	..	—	52	..	0
19	..	—	53	..	0
20	..	—	54	..	—
21	..	27	55	..	—
22	..	54	56	..	35
23	..	5	57	..	0
24	..	10	58	..	0
25	..	20	59	..	—
26	..	40	60	..	—
27	..	—	61	..	—
28	..	81	62	..	—
29	..	—	63	..	—
30	..	15	64	..	0
31	..	30	65	..	—
32	..	—	66	..	49
33	..	7			

Col. 1 of Table 1 contains zero and the nine digits, and col. 2 of the same Table the corresponding *simple index numbers.* Col. 1 of Table 2 sets forth all *partial products* (a term applied to the product of any two units), while col. 2 contains the corresponding *compound index numbers*. The relation between the index numbers is such that the sum of the simple index numbers of any two units is equal to the compound index number of their product. Table 3 is really a re-arrangement of Table 2, the numbers 0 to 66 (representing 67 divisions on the index) being placed in col. 1, and in col. 2, opposite to each number in col. 1 which is a compound index number, is placed the corresponding simple product.

Now, to take a very simple example, suppose the machine is supplied with a formula-paper designed to cause it to evaluate x for given values of a, b, c, and d, in the equation $ab + cd = x$, and suppose we wish to find the value of x in the particular case where $a = 9247$, $b = 8132$, $c = 21893$, and $d = 823$.

The four given numbers are first transferred to the machine by the key-board hereafter mentioned; and the formula-paper causes them to be inscribed in four shuttles. As the shuttles of the inner and outer co-axial shuttle-boxes are numbered consecutively, we may suppose the given values of a and c to be inscribed in the first and second shuttles respectively of the inner box, and of b and d in the first and second shuttles respectively of the outer box; but it is important to remember that it is a function of the formula-paper to select the shuttles to receive the Variables, as well as the shuttles to be operated on, so that (except under certain special circumstances, which arise only in more complicated formulæ) any given formula-paper always selects the same shuttles in the same sequence and manner, whatever be the values of the Variables. The magnitude of a Variable only effects the type carried by its shuttle, and in no way influences the movements of the shuttle as a whole.

The machine, guided by the formula-paper, now causes the shuttle-boxes to rotate until the first shuttles of both inner and outer boxes come opposite to a shuttle-race. The two shuttles are then drawn along the race to a position near the index; and certain slides are released, which move forward until stopped by striking the type carried by the outer shuttle. The slides in question will then have moved distances corresponding to the simple index numbers of the corresponding digits of the Variables b. In the particular case under consideration, the first four slides will therefore move 3, 0, 7, and 1 units respectively, the remainder of the slides indicating zero by moving 50 units (see Table 1). Another slide moves *in the opposite direction* until stopped by the first type of the inner shuttle, making a movement proportional to the simple index number of the first digit of the multiplier a—in this case 14. As the index is attached to the last-mentioned slide, and partakes of its motion, the *relative* displacements of

Percy E. Ludgate, Scientific Proceedings of the Royal Dublin Society, Vol. 12, No. 9, pp. 77–91, 1909. © 1909, RDS

On a proposed Analytical Machine

the index and each of the four slides are respectively $3 + 14$, $0 + 14$, $7 + 14$, and $1 + 14$ units (that is 17, 14, 21, and 15 units), so that pointers attached to the four slides, which normally point to zero on the index, will now point respectively to the 17th, 14th, 21st, and 15th divisions of the index. Consulting Table 3, we find that these divisions correspond to the partial products 72, 9, 27, and 18. In the index the partial products are expressed mechanically by movable blades placed at the intervals shown in column 2 of the third table. Now, the duration of the first movement of any blade is as the unit figure of the partial product which it represents, so that the movements of the blades concerned in the present case will be as the numbers 2, 9, 7, and 8, which movements are conveyed by the pointers to the mill, causing it to register the number 2978. A carriage near the index now moves one step to effect multiplication by 10, and then the blades partake of a second movement, this time transferring the tens' figures of the partial products (*i.e.* 7, 0, 2, and 1) to the mill, which completes the addition of the units' and tens' figures thus—

$$2978$$
$$7021$$
$$\overline{}$$
$$73188$$

—the result being the product of the multiplicand b by the first digit of the multiplier a. After this the index makes a rapid reciprocating movement, bringing its slide into contact with the second type of the inner shuttle (which represents the figure 2 in the quantity a), and the process just described is repeated for this and the subsequent figures of the multiplier a until the whole product ab is found. The shuttles are afterwards replaced in the shuttle-boxes, the latter being then rotated until the second shuttles of both boxes are opposite to the shuttle-race. These shuttles are brought to the index, as in the former case, and the product of their Variables (21893×823) is obtained, which, being added to the previous product (that product having been purposely retained in the mill), gives the required value of x. It may be mentioned that the position of the decimal point in a product is determined by special mechanism which is independent of both mill and index.

Most of the movements mentioned above, as well as many others, are derived from a set of cams placed on a common shaft parallel to the driving-shaft; and all movements so derived are under the control of the formula-paper.

The ordinals in Table 1 are not mathematically important, but refer to special mechanism which cannot be described in this paper, and are included in the tables merely to render them complete.

The sum of two products is obtained by retaining the first product in the mill until the second product is found—the mill will then indicate their sum. By reversing the direction of rotation of the mill before the second product is obtained, the difference of the products results. Consequently, by making the multiplier unity in each case, simple addition and subtraction may be performed.

In designing a calculating machine it is a matter of peculiar difficulty and of great importance to provide for the expeditious carrying of tens. In most machines the carryings are performed in rapid succession; but Babbage invented an apparatus (of which I have been unable to ascertain the details) by means of which the machine could "foresee" the carryings and act on the foresight. After several years' work on the problem, I have devised a method in which the carrying is practically in complete mechanical independence of the adding process, so that the two movements proceed simultaneously. By my method the sum of m numbers of n figures would take $9m + n$ units of time. In finding the product of two numbers of twenty figures each, forty additions are required (the units' and tens' figures of the partial products being added separately). Substituting the values 40 and 20 for m and n, we get $9 \times 40 + 20 = 380$, or $9\frac{1}{2}$ time-units for each addition—the time-unit being the period required to move a figure-wheel through $\frac{1}{10}$ revolution. With Variables of 20 figures each the quantity n has a constant value of 20, which is the number of units of time required by the machine to execute any carrying which has not been performed at the conclusion of an indefinite number of additions. Now, if the carryings were performed in succession, the time required could not be less than $9 + n$, or 29 units for each addition, and is, in practice, considerably greater.[1]

In ordinary calculating machines division is accomplished by repeated subtractions of the divisor from the dividend. The divisor is subtracted from the figures of the dividend representing the higher powers of ten until the remainder is less than the divisor. The divisor is then moved one place to the right, and the subtraction proceeds as before. The number of subtractions performed in each case denotes the corresponding figure of the quotient. This is a very simple and convenient method for ordinary calculating machines; but it scarcely meets the requirements of an Analytical Machine. At the same time, it must be observed that Babbage used this method, but found it gave rise to many mechanical complications.

---

[1] For further notes on the problem of the carrying of tens, see C. Babbage: "Passages from the Life of a Philosopher," p. 114, &c.

Percy E. Ludgate, Scientific Proceedings of the Royal Dublin Society, Vol. 12, No. 9, pp. 77–91, 1909. © 1909, RDS

On a proposed Analytical Machine

Percy E. Ludgate, Scientific Proceedings of the Royal Dublin Society, Vol. 12, No. 9, pp. 77–91, 1909. © 1909, RDS

On a proposed Analytical Machine

My method of dividing is based on quite different principles, and to explain it I must assume that the machine can multiply, add, or subtract any of its Variables; or, in other words, that a formula-paper can be prepared which could direct the machine to evaluate any specified function (which does not contain the sign of division or its equivalent) for given values of its variables.

Suppose, then, we wish to find the value of $\frac{p}{q}$ for particular values of p and q which have been communicated to the machine. Let the first three figures of q be represented by f, and let A be the reciprocal of f, where A is expressed as a decimal of 20 figures. Multiplying the numerator and denominator of the fraction by A, we have $\frac{Ap}{Aq}$, where Aq must give a number of the form 100 . . . because $Aq = \frac{q}{f}$. On placing the decimal point after the unit, we have unity plus a small decimal. Represent this decimal by x: then—

$$\frac{p}{q} = \frac{Ap}{1 + x} \quad \text{or} \quad Ap\,(1 + x)^{-1}$$

Expanding by the binomial theorem—

(1) $\quad \frac{p}{q} = Ap\,(1 - x + x^2 - x^3 + x^4 - x^5 + \&\text{c.}),$

or

(2) $\quad \frac{p}{q} = Ap\,(1 - x)\,(1 + x^2)\,(1 + x^4)\,(1 + x^8),\ \&\text{c.}$

The series (1) converges rapidly, and by finding the sum as far as x^{10} we obtain the correct result to at least twenty figures; whilst the expression (2 gives the result correctly to at least thirty figures. The position of the decimal point in the quotient is determined independently of these formulæ. As the quantity A must be the reciprocal of one of the numbers 100 to 999, it has 900 possible values. The machine must, therefore, have the power of selecting the proper value for the quantity A, and of applying that value in accordance with the formula. For this purpose the 900 values of A are stored in a cylinder—the individual figures being indicated by holes of from one to nine units deep in its periphery. When division is to be performed, this cylinder is rotated, by a simple device, until the number A (represented on the cylinder by a row of holes), which is the reciprocal of the first three figures of the divisor, comes opposite to a set of rods. These rods then transfer that number to the proper shuttle, whence it becomes an ordinary Variable, and is used in accordance with the formula. It is not necessary that every time the process of division is required the dividing formula should be worked out in detail in the formula-paper. To obviate the necessity of so doing the machine is provided with a special permanent *dividing cylinder*, on which this formula is represented in the proper notation of perforations. When the arrangement of perforations on the formula-paper indicates that division is to be performed, and the Variables which are to constitute divisor and dividend, the formula-paper then allows the dividing cylinder to usurp its functions until that cylinder has caused the machine to complete the division.

It will be observed that, in order to carry out the process of division, the machine is provided with a small table of numbers (the numbers A) which it is able to consult and apply in the proper way. I have extended this system to the logarithmic series, in order to give to that series a considerable convergency; and I have also introduced a *logarithmic cylinder* which has the power of working out the logarithmic formula, just as the dividing cylinder directs the dividing process. This system of auxiliary cylinders and tables for special formulæ may be indefinitely extended.

The machine prints all results, and, if required, the data, and any noteworthy values which may transpire during the calculation. It may be mentioned, too, that the machine may be caused to calculate and print, quite automatically, a table of values—such, for instance, as a table of logs, sines, squares, &c. It has also the power of recording its results by a system of perforations on a sheet of paper, so that when such a *number-paper* (as it may be called) is replaced in the machine, the latter can "read" the numbers indicated thereon, and inscribe them in the shuttles reserved for the purpose.

Among other powers with which the machine is endowed is that of changing from one formula to another as desired, or in accordance with a given mathematical law. It follows that the machine need never be idle; for it can be set to tabulate successive values of any function, while the work of the tabulation can be suspended at any time to allow of the determination by it of one or more results of greater importance or urgency. It can also "feel" for particular events in the progress of its work—such, for instance, as a change of sign in the value of a function, or its approach to zero or infinity; and it can make any pre-arranged change in its procedure, when any such event occurs. Babbage dwells on these and similar points, and explains their bearing on the automatic solution (by approximation) of an equation of the nth degree;[1] but I have not been able to ascertain whether his way of attaining these results has or has not any resemblance to my method of so doing.

---

[1] C. Babbage: "Passages from the Life of a Philosopher," p. 131.

Percy E. Ludgate, Scientific Proceedings of the Royal Dublin Society, Vol. 12, No. 9, pp. 77–91, 1909. © 1909, RDS

On a proposed Analytical Machine

On a proposed Analytical Machine

Percy E. Ludgate, Scientific Proceedings of the Royal Dublin Society, Vol. 12, No. 9, pp. 77–91, 1909. © 1909, RDS

The Analytical Machine is under the control of two key-boards, and in this respect differs from Babbage's Engine. The upper key-board has ten keys (numbered 0 to 9), and is a means by which numbers are communicated to the machine. It can therefore undertake the work of the number-paper previously mentioned. The lower key-board can be used to control the working of the machine, in which case it performs the work of a formula-paper. The key-boards are intended for use when the nature of the calculation does not warrant the preparation of a formula-paper or a number-paper, or when their use is not convenient. An interesting illustration of the use of the lower key-board is furnished by a case in which a person is desirous of solving a number of triangles (say) of which he knows the dimensions of the sides, but has not the requisite formula-paper for the purpose. His best plan is to put a plain sheet of paper in the controlling apparatus, and on communicating to the machine the known dimensions of one of the triangles by means of the upper key-board, to guide the machine by means of the lower key-board to solve the triangle in accordance with the usual rule. The manipulations of the lower key-board will be recorded on the paper, which can then be used as a formula-paper to cause the machine automatically to solve the remaining triangles. He can communicate to the machine the dimensions of these triangles individually by means of the upper key-board; or he may, if he prefers so doing, tabulate the dimensions in a number-paper, from which the machine will read them of its own accord. The machine is therefore able to "remember," as it were, a mathematical rule; and having once been shown how to perform a certain calculation, it can perform any similar calculation automatically so long as the same paper remains in the machine.

It must be clearly understood that the machine is designed to be quite automatic in its action, so that a person almost entirely ignorant of mathematics could use it, in some respects, as successfully as the ablest mathematician. Suppose such a person desired to calculate the cosine of an angle, he obtains the correct result by inserting the formula-paper bearing the correct label, depressing the proper number-keys in succession to indicate the magnitude of the angle, and starting the machine, though he may be quite unaware of the definition, nature, or properties of a cosine.

While the machine is in use its central shaft must be maintained at an approximately uniform rate of rotation—a small motor might be used for this purpose. It is calculated that a velocity of three revolutions per second would be safe; and such a velocity would ensure the multiplication of any two Variables of twenty figures each in about 10 seconds, and their addition or subtraction in about 3 seconds. The time taken to divide one Variable by another depends on the degree of convergency of the series derived from the divisor, but $1\frac{1}{2}$ minutes may be taken as the probable maximum. When constructing a formula-paper, due regard should therefore be had to the relatively long time required to accomplish the routine of division; and it will, no doubt, be found advisable to use this process as sparingly as possible. The determination of the logarithm of any number would take 2 minutes, while the evaluation of a^n (for any value of n) by the expotential theorem, should not require more than $1\frac{1}{2}$ minutes longer— all results being of twenty figures.[1]

The machine, as at present designed, would be about 26 inches long, 24 inches broad, and 20 inches high; and it would therefore be of a portable size. Of the exact dimensions of Babbage's Engine I have no information; but evidently it was to have been a ponderous piece of machinery, measuring many feet in each direction. The relatively large size of this engine is doubtless due partly to its being designed to accommodate the large number of one thousand Variables of fifty figures each, but more especially to the fact that the Variables were to have been stored on columns of wheels, which, while of considerable bulk in themselves, necessitated somewhat intricate gearing arrangements to control their movements. Again, Babbage's method of multiplying by repeated additions, and of dividing by repeated subtractions, though from a mathematical point of view very simple, gave rise to very many mechanical complications.[2]

To explain the power and scope of an Analytical Machine or Engine, I cannot do better than quote the words of the Countess of Lovelace: "There is no finite line of demarcation which limits the powers of the Analytical Engine. These powers are coextensive with the knowledge of the laws of analysis itself, and need be bounded only by our acquaintance with the latter. Indeed, we may consider the engine as the material and mechanical representative of analysis, and that our actual working powers in this department of human study will be enabled more effectually than heretofore to keep pace with our theoretical knowledge of its principles and laws, through the complete control which the engine gives us over the executive manipulations of algebraical and numerical symbols."[3]

A Committee of the British Association which was appointed to report

---

[1] The times given include that required for the selection of the Variables to be operated on.
[2] See Report Brit. Assoc., 1878, p. 100.
[3] R. Taylor's "Scientific Memoirs," 1843, vol. iii., p. 696.

on Babbage's Engine stated that, " apart from the question of its saving labour in operations now possible, we think the existence of such an instrument would place within reach much which, if not actually impossible, has been too close to the limits of human skill and endurance to be practically available."[1]

In conclusion, I would observe that of the very numerous branches of pure and applied science which are dependent for their development, record, or application on the dominant science of mathematics, there is not one of which the progress would not be accelerated, and the pursuit would not be facilitated, by the complete command over the numerical interpretation of abstract mathematical expressions, and the relief from the time-consuming drudgery of computation, which the scientist would secure through the existence of machinery capable of performing the most tedious and complex calculations with expedition, automatism, and precision.

---

[1] Report Brit. Assoc., 1878, p. 101.

Percy E. Ludgate, Scientific Proceedings of the Royal Dublin Society, Vol. 12, No. 9, pp. 77–91, 1909. © 1909, RDS

On a proposed Analytical Machine

On a proposed Analytical Machine

Percy E. Ludgate, Scientific Proceedings of the Royal Dublin Society, Vol. 12, No. 9, pp. 77–91, 1909. © 1909, RDS

Royal Dublin Society.

FOUNDED, A.D. 1731.   INCORPORATED, 1749.

EVENING SCIENTIFIC MEETINGS.

The Scientific Meetings of the Society are held alternately at 4.30 p.m. and 8 p.m. on the third Tuesday of every month of the Session (November to June).

Authors desiring to read Papers before the Society are requested to forward their Communications to the Registrar of the Royal Dublin Society *at least* ten days prior to each Meeting, as no Paper can be set down for reading until examined and approved by the Science Committee.

The copyright of Papers read becomes the property of the Society, and such as are considered suitable for the purpose will be printed with the least possible delay. Authors are requested to hand in their MS. and necessary Illustrations in a complete form, and ready for transmission to the Editor.

DUBLIN : PRINTED AT THE UNIVERSITY PRESS BY PONSONBY AND GIBBS.

On a proposed Analytical Machine

Percy E. Ludgate, Scientific Proceedings of the Royal Dublin Society, Vol. 12, No. 9, pp. 77–91, 1909. © 1909, RDS

From: *Percy E. Ludgate Prize in Computer Science*, TCD-SCSS-X.20121208.002, 11th October, 2019. © 2019, The John Gabriel Byrne Computer Science Collection

Extracts of Percy Ludgate's Genealogy

Ancestors of Percy Edwin Ludgate

11 October 2019

First Generation

1. **Percy Edwin Ludgate** was born on 2 Aug 1883 at Townsend Street in Skibbereen, Cork, Ireland.[1] Father Michael Ludgate, mother Mary Ludgate formerly McMahon. Fathers profession Pensioner.
Registered 20th September, informant Mother He was educated on 15 Sep 1890 at St George's Infants in Dublin, Ireland.[2] Age 7. Living at 28 Foster Terrace. Father a Shorthand Teacher. Member of the Established Church. He was educated on 31 Mar 1891–31 Mar 1892 at St George's Infants in Dublin, Ireland.[3] Established Church, father a teacher. Living at 28 Foster Terrace, Dublin. Transferred from Taft? Hall. On 4 Nov 1898 Percy was a Civil Service Temporary Boy Copyist (New Class) in Dublin, Ireland.[4] He appeared in the census on 31 Mar 1901 at 30 Dargle Road in Glasnevin, Dublin, Ireland.[5] Age 17 Church of Ireland born County Cork occupation Civil Servant National Education Office (Boy Copyist) Read and write Not Married
Also present mother Mary (age 60) and brother Alfred (age 19)
On 13 Mar 1903 he was a Civil Service Open Competitive Examination for situations as Assistant Clerks (Abstractors) in Dublin, Ireland.[6] He was the **top Irish candidate** being placed nineteenth in the Order of Merit. On 18 Oct 1904 Percy was a Civil Service Open Competitive Examinations for Clerkships in the Second Division of the Civil Service in Dublin, Ireland.[7] Percy competed successfully for this more senior graded clerkship but failed the medical examination. Consequently his certificate was not issued On 20 Feb 1905 he was a "Case of Mr. Percy Ludgate – Irish Civil Service" was raised during Questions in the House in London, England.[8] Mr. T. HARRINGTON (Dublin Harbour)

I beg to ask the Secretary to the Treasury whether he is aware that Mr. Percy E. Ludgate, of Drumcondra, passed the Civil Service examination for assistant clerkship,abstractor class, in October, 1903, and was medically examined by the physician selected by the Civil Service Commissioners and declared fit for the service;that, without having received an appointment as assistant clerk, he competed successfully for a second division clerkship, but failed to satisfy the examining physician as to his fitness; that, in consequence of the latter medical examination, Mr. Ludgate's certificate of qualification given in the former case has been cancelled, thus penalising him for one appointment by reason of his success in securing another; and, if so, whether he will take steps to have him medically examined with the view to his securing one or other of the above appointments, and will he say whether there is any difference in the medical standard of qualification required for assistant clerkship or second-division clerkship.

THE FINANCIAL SECRETARY OF THE TREASURY (Mr. VICTOR CAVENDISH,) Derbyshire, W.

The facts are substantially as stated in the hon. Member's Question, except that Mr. Ludgate's certificate as assistant clerk was never granted, and therefore was not cancelled; and that the date of the examination referred to was February, 1903, and not October, 1903. As nearly a year had elapsed since Mr. Ludgate's medical examination for an assistant clerkship, it was necessary to re-examine him before issuing him a certificate for a second-division clerkship. The result of the medical examination proving unsatisfactory the Civil Service Commissioners were unable to grant certificate for either position. The medical requirements are practically the same in both cases. On 28 Apr 1909 he was a "On a Proposed Analytical Machine at Royal Dublin Society in Dublin, Ireland.[9] I propose to give in this paper a short account of the results of about six years' work, undertaken by me with the object of designing machinery capable of performing calculations, however intricate or laborious, without the immediate guidance of the human intellect. Percy appeared in the census on 2 Apr 1911 at 30 Dargle Road in Glasnevin, Dublin, Ireland.[10] Age 27 Church of Ireland born Co Cork Commercial Clerk (Corn Merchant)
Living with mother Mary (age 70) and brother Alfred (age 29)
In 1912 he was an Automatic Calculating Machines. By P. E. LUDGATE. at Fifth International Congress of

---

1. , , Skibbereen Registration District Page 67 No 335, ; , .
2. , "Ireland National School Registers," Roll Number 11624, ; , .
3. , "Ireland National School Registers," Ireland National School Registers No 11624.
4. *Newspapers*, , London Gazette. 4th November 1898 Pages 6454 & 6455.
5. Census, ; Census Ireland 1901 Household Return.
6. *Newspapers*, , London Gazette 17th March 1903 page 1779 & Weekly Irish Times 21st March 1903 (London Correspondence).
7. *Newspapers*, , London Gazette 23rd August 1904 pages 5419 and 5420.
8. , Hansard 20th February 1905 Vol 141 cc619-20.
9. , Scientific Proceedings Royal Dublin Society 12,9 (1909) pp 77-91.
10. Census, , .

Mathematicians in Cambridge, Cambridgeshire, England.[11] He was educated Accountants Examinations in Jun 1917 in London, England.[12] Results for the June Examination of the Corporation of Accountants Limited. In the final all passed, Percy E. Ludgate, William Codd and Thomas sanderson, Dublin taking Honours Percy Grant in Perpetuity of Burial in Mount Jerome Cemetery on 6 Dec 1921 at Mount Jerome Cemetery in Dublin, Ireland.[13] Plot of Ground "A" measuring 6' 6" x 2' 6" sub division 412 granted to Percy E Ludgate for £6-0-0.

16113 is the registered number of the grave distinguished by A29-410

Buried

Alfred Ernest Ludgate

Fredrick Ludgate

Percy E Ludgate

Alice Emily Ludgate

Mary Ann Ludgate

Full

He died on 16 Oct 1922 at the age of 39 at 30 Dargle Road in Glasnevin, Dublin, Ireland.[14] Bachelor age 38, profession Accountant. Cause of death Catarhal Pneumonia 21 Days certified. Informant Violet E. Ludgate, Niece, in attendance, of 1 Tolka Villas, Richmond Road. He had his estate probated on 23 Jan 1923 at 30 Dargle Road in Glasnevin, Dublin, Ireland.[15] Probate granted at Dublin to Alfred E. Ludgate, Accountant, Effects £885 7s 4d. Percy had his estate probated on 12 Feb 1923 at Court of Chancery in London, England.[16] Probate Dublin to Alfred Ernest Ludgate, accountant. Effects £192 in England. Sealed London 12 February (1923)

11. , NapierTercentenary Handbook1914-CalculatingMachines-Whipple and Ludgate.

12. *Newspapers*, Sept 10, 1917 Birmingham Daily Gazette, Birmingham Evening Despatch. Sept 11, 1917 Yorkshire Post and Leeds Intelligencer, British Library Newspaper Collection.

13. , Mount Jerome Cemetery.

14. , , Registration District Dublin, Finglass and Glasnevin 04388107 No 215, .

15. , Dublin Callendar Court of Chancery, .

16. , Will Calendars, Court of Chancery, London.

Second Generation

2. **Michael Edward Ludgate** was christened on 8 Feb 1840 in Kilshannig by Mallow, Cork, Ireland.[17] Place described as Kilshannig by Mallow. Parents Robert and Susanna Ludgate He served in the military North Cork Militia on 15 Sep 1857–9 Jun 1857 in Cork, Ireland.[18] 15 Sept 1857 Recruit born Mallow
1 November 1857 Promoted Corporal with David and Robert Ludgate
1 April 1858 Regiment moves to Hyde, England
8 June 1858 Volunteers for 21st Regiment of Foot
 He served in the military 21st Regiment of Foot on 9 Jun 1858–1 Oct 1861.[19] WO 12/3851 Hythe 21st Foot 14-Jun 1858 Voluntary enlisted from Militia
WO 12/3851 Shincliffe 21st Foot 30-Jun 1858 Muster as Corporal
WO 12/3851 Hythe 21st Foot 30-Jun 1858 Voucher no 14 : 5 Volunteers from North Cork Volunteers
WO 12/3851 Hythe 21st Foot 30-Sep 1858 Muster as Corporal
WO 12/3854 Chatham 21st Foot 27-Apr 1861 Muster Sergeant
WO 12/3853 Aldershot 21st Foot 30-Jun 1861 Muster as Sergeant
WO 12/3854 Hythe 21st Foot 30-Sep 1861 No longer with Regiment
WO 12/3854 Hythe 21st Foot 01-Oct 1861 Sent to Winchester School of Musketry
WO 12/3854 Hythe 21st Foot 01-Oct 1861 Discharged 21st Foot
WO 12/3854 Hythe 21st Foot 01-Oct 1861 Voucher No3 Authority for Discharge of Sgt M Ludgate
 Michael appeared in the census in Apr 1861 at Shorncliff Camp in Cheriton, Kent, England.[20] Relationship Quartered In Camp At Shorncliff
Marital status Single
Gender Male
Age 22
Birth year 1839
Occupation Serjeant 2/21 Regiment
Birth place (other) Ireland
Parish Cheriton
County Kent
Country England
Parliamentary borough Hythe
Registration district Elham
 He served in the military 60th Regiment of Foot Attached to School of Musketry on 1 Oct 1861–19 Dec 1876.[21] WO 12/7031 Chatham 60th Foot 01-Apr 30-Jun 1869 Married establishment since 7-11-1869. Wife Mary 3 Children 10, 4-2,0-2
WO 12/7027 Bellary, India 60th Foot 1871 1871 Index "Sergeant Instructor in Musketry"
WO12 /7028 Isthmus (Aden) 60th Foot 30-Apr 30-Jun 1872 Muster Sargeant Instructor of Musketry
WO12 /7028 Aldershot 60th Foot 24-Dec 31-Mar 1873 Muster Sargeant Instructor of Musketry
WO12 /7029 Shorncliffe Camp 60th Foot 09-Apr 30-Jun 1873 Muster Sargeant Instructor of Musketry
WO 12/7031 Chatham 60th Foot 01-Apr 30-Jun 1875 Muster Sergeant Instructor of Musketry, on attatchment.
WO 12/7031 Chatham 60th Foot 01-Jul 30-Sep 1875 Muster Sergeant Instructor of Musketry, on attatchment.
WO 12/7032 Winchester 60th Foot 01-Apr 23-May 1876 Appointed Master Sergeant Instructor in Musketry 24/5-30/6
WO 12/7032 Winchester 60th Foot 01-Apr 30-Jun 1876 Married establishment since 7-11-1869. Wife Mary 3 Children 10-9, 4-11,0-11
WO 12/7032 Gravesend 60th Foot 01-Jul 30-Sep 1876 Quarter Master Sergeant. Payment permitted in the 76th Attached to the 29th Foot.
WO 12/7032 Gravesend 60th Foot 01-Jul 30-Sep 1876 Married establishment since 7-11-1869. Wife Mary 3 Children 11, 5-2,1-3
WO 12/7033 Winchester 60th Foot 24-Oct 19-Dec 1876 Married establishment since 7-11-1869. Wife Mary 3 Children 11-2, 5-8,1-5
WO 12/7032 Winchester 60th Foot 19-Dec 1876 Discharge Modified pension. Wife and 3 children 11,5,1 Tickets Winchester - Bristol, Bristol Cork

17. , Familysearch Film FHL 874437 Item 7.
18. , National Archives UK WO13 /2723.
19. , National Archives UK WO 12/ 3851 /3853 /3854.
20. Census, , .
21. , National Archives WO 12 /7028-7033.

He served in the military Pension Payment in 1882–1883 in Skibbereen, Cork, Ireland.[22] On 24 Apr 1888 Michael was a Clerk.[23] In 1891–1897 he was a Ludgate, Michael Edward, son and daughter, teacher of shorthand at 28 Foster Terrace in Dublin, Ireland.[24] Fredrick and Augusta, teachers of shorthand He Non payment of debt on 8 Sep 1899–19 Oct 1899 at Kilmainham Prison in Dublin, Ireland.[25] Prison Number: 1076

Names: Michael Ed Ludgate (Debtors) (Army Pensioner)

Description:

 Age: 60

 Height Ft & Inch: 5' 11¾"

 Hair/Eyes: Grey/Blue

 Complexion: Fresh

Marks on Person: Bald and Moles on Chest and right side of stomach sore with left part of back. Left eye impaired Lost his upper teeth except three.

Weight on Admission/Discharge: 179/

Where Born: Mallow,Co Cork

Last Residence: Balbriggan

Trade or Occupation: Nil

Religion: C.I.

Degree of Education: R+W

When committed: Sept. 8

Offence: Non-paymentof Debt

By whom committed: TheHon Mr Justice Kenny, High Court of Justice, Ireland 16.8.99

Sentence: 6 Weeks or £5:2:0

Fine, Bail or Hard Labour: impt (Imprisonment)

Expiration of Sentence: 22 Sept 1899 On 27 Aug 1900 Michael was a Traveller in Dublin, Ireland.[26] From Marriage certificate Frederic Ludgate He appeared in the census in Apr 1901 at 14 Quay Street in Balbriggan, Dublin, Ireland.[27] Episcopalian Church of Ireland age 61 Pensioner from 60th Rifles, born Co Cork He appeared in the census in Apr 1911 at 6 John Street in Omagh, Northern Ireland.[28] Surname Forename Age Sex Relation to head Religion

Milligan William John 42 Male Head of Family Methodist

Milligan Lizzie   45 Female Wife Methodist

Milligan Emma Janietta 16 Female Daughter Methodist

Milligan Louisa Madaline 15 Female Daughter Methodist

Milligan Fredrick James 13 Male Son Methodist

Ludgate Edward 65 Male Boarder Episcopalian Church of Ireland

Henry John 27 Male Boarder R Catholic

Lindsay Maggie 22 Female Servant Episcopalian Church of Ireland

 Michael retired in 1919 in Belfast, Antrim, Northern Ireland.[29] Name Michael Edward

Surname LUDGATE

Father Robert LUDGATE

Mother Susan WILLIS

Townland Kilshannig Upr. or Lr.

Parish Rathcormack

Barony Barrymore

County Cork

Age of Applicant 79

1841 Census

1851 Census N/T

Observation Robert LUDGATE farmer. Try Kilshanny, Brigown, Condond and Clangibbon, Cork - N/T

Ireland Genealogy

Source Film ID 0993092

22. , National Archives UK WO22.

23. , , , .

24. , Thom's Directory, 1891 Dublin Street Directory 1891-1897.

25. , FMP DUBLIN-KILMAINHAM PRISON GENERAL REGISTER 1898-1903 Bk 1/10/21 Item 2.

26. , Marrige Certificate of Fredrick Ludgate and Alice Walsh 3q 1900.

27. Census, , .

28. Ibid.

29. , Pension Applications Film ID 0993092 Ireland Gen ID 4705.

Ireland Gen ID 4705

He died Cause of Death Bronchitis on 26 Jan 1923 at Union Infirmary in Belfast, Antrim, Northern Ireland.[30] Age 82 Army Pensioner Single Episcopalian He was buried on 30 Jan 1923 at Belfast City Cemetery in Belfast, Antrim, Northern Ireland.[31] Burial Managed by Mr Morton of the Belfast Union Mary Ann McMahon and Michael Edward Ludgate were married on 15 Aug 1863 in Winchester, Hampshire, England.[32] Michael Ludgate Age 23 Musketry Instructor
Father's name(s) Robert Ludgate Farmer
Spouse Mary McMahon age 22 School Mistress
Father's name(s) Thomas McMahon Soldier

3. **Mary Ann McMahon** was born on 19 Nov 1840 in Iden, Sussex, England, United Kingdom.[33] Father Private in the Royal Sappers and Miners She was christened on 20 Dec 1840 at All Saints in Iden, Sussex, England.[34] She appeared in the census in Apr 1841 at East Street in Plumstead, Kent, England.[35] Thomas McMahan Male 35 1806 Ireland
Francis McMahan Female 30 1811 Ireland
Augusta McMahan Female 7 1834 Kent, England
Thomas McMahan Female 5 1836 Kent, England
Auther McMahan Male 3 1838 Kent, England
Mary Ann McMahan Female 0 1841
-
Mary appeared in the household of Frances "Fanny" Reed in the census in 1851 at Royal Hospital Chelsea in Chelsea, Middlesex, England.[36] She appeared in the census in 1851 at Royal Hospital Chelsea in Chelsea, Middlesex, England.[37] Age 10 Scholar b Rye, Sussex. Living with Mother She appeared in the census in Apr 1861 at Royal Hospital Chelsea in Chelsea, Middlesex, England.[38] Relationship Nurses Daughter
Marital status Unmarried
Gender Female
Age 20
Birth year 1841
Occupation Dressmaker
Birth town Rye
Birth county Sussex
Living at this address[39] Living at this address[40] Mary died on 22 Aug 1936 at the age of 95 at 2 Belvidere Ave in Dublin, Ireland.[41] Died St.Kevin's Hospital, Dublin. Cause of Death Senile Decay Cardiac Arrest. Informant Hospital

Michael Edward Ludgate and Mary Ann McMahon had the following children:

    i. **Arthur Edward Ludgate** was born on 14 Jul 1864 at Barracks in Winchester, Hampshire, England.[42] He died Pneumonia on 10 Dec 1864 at the age of 0 at Barracks in Winchester, Hampshire, England.[43]
    ii. **Thomas Edward Ludgate**[44] was born on 23 Sep 1865 at Barracks in Winchester, Hampshire,

30. , Belfast City Cemetery Application for Internmant 58375.
31. Ibid.
32. , , , ; , .
33. Ibid.
34. , Familysearch FHL Film 1067258.
35. Census, , .
36. Ibid.
37. Ibid.
38. Ibid.
39. Census, , Census Ireland 1901 Household Return.
40. Census, , .
41. , , , , .
42. Ibid.
43. Ibid.
44. Ibid.

Ancestors of Percy Edwin Ludgate

England.[45] He died in Mar Q 1951 at the age of 85 in Lewes, Sussex, England.[46]

iii. **Walter Samuel Ludgate** was born on 13 Nov 1867 at Barracks in Winchester, Hampshire, England.[47] He died Hydrocephalis on 21 Nov 1868 at the age of 1 at Barracks in Winchester, Hampshire, England.[48]

iv. **Albert William Ludgate** was born on 22 Dec 1868 at Barracks in Winchester, Hampshire, England.[49] He died on 24 Jun 1870 at the age of 1 at Bellary in Madras, India.[50] Hydrocephalus. Buried 25th June 1871

v. **Augusta Ludgate** was born on 3 Mar 1871 at Bellary in Madras, India.[51]
She was christened on 4 May 1871 at Bellary in Madras, India.[52] She appeared in the census in 1901 at 1 Blackburne Place in Liverpool, Lancashire, England.[53] Age 30 "In Charge Visitor" Marie Futty. A deaconess (missionary) born India. Augusta appeared in the census in Sep 1939 at London County Council Mental Hospital in Abbots Langley, Hertfordshire, England.[54] b 1874 Female Incapacitated Single
She died on 30 Dec 1954 at the age of 83 at Leavesden Hospital, Abbots Langley in Watford, Hertfordshire, England.[55] Age on certificate 80, formerly a typist
a) Broncho Pneumonia
b) Myocardial generation
c) Cystitis

vi. **Fredrick Ludgate** was born on 17 Jun 1875 at Milton Barracks in Gravesend, Kent, England,.[56] Father Sergeant Instructor of Musketry He died on 2 Dec 1921 at the age of 46 at 1 Tolka Villas, Richmond Road, Dublin in Dublin, Ireland.[57] Married , 45 years a clerk. Case of death Pulmonary Tuberculosis, informant Alice E. Ludgate of 1 Tolka Villas He was buried on 6 Dec 1921 at Mount Jerome Cemetery in Dublin, Ireland.[58] Grave 39 A412. "Accountant" Alfred Ludgate

vii. **Alfred Ludgate** was born on 1 Apr 1881 at Townsend Street in Skibbereen, Cork, Ireland.[59] Informant Michael Ludgate Living at this address[60] Living at this address[61] He lived Electoral Roll at 30 Dargle Road in Glasnevin, Dublin, Ireland in 1915. House and small garden. Rated occupier £13. He died on 10 Sep 1953 at the age of 72 at Royal Victoria Eye and Ear Hospital in Dublin, Ireland.[62] of 10 Emmet Street, N.C.R. single, age 72 a clerk. Cause of death Larynx (1 year) Heart failure 1 day, certified.

1 viii. **Percy Edwin Ludgate**, born 2 Aug 1883, Skibbereen, Cork, Ireland; died 16 Oct 1922, Glasnevin, Dublin, Ireland.

45. Ibid.
46. Ibid.
47. Ibid.
48. Ibid.
49. Ibid.
50. Ibid.
51. Ibid.
52. Ibid.
53. Census, , .
54. , FMP 1939 Register RG101/1674G/006/15.
55. , , , .
56. Ibid.
57. Ibid.
58. , Mount Jerome Cemetery Register 1519.
59. , , , .
60. Census, , Census Ireland 1901 Household Return.
61. Census, , .
62. , , , .

Ancestors of Percy Edwin Ludgate

Third Generation

4. **Robert Ludgate** was christened on 16 Aug 1802 at Scarragh in Kilshannig by Mallow, Cork, Ireland.[63] He lived at Scarragh in Kilshannig by Mallow, Cork, Ireland in 1852.[64] Susanna Willis and Robert Ludgate were married in 1837 at Diocese of Cork & Ross in Cork, Ireland.[65]

5. **Susanna Willis** was born about 1814. She died on 15 Aug 1894 at the age of 80 at Incurable Home in Cork, Cork, Ireland.[66] Widow age 80, no occupation, Coronary H??? one year, Exhaustion. Informant H McLaine, occupier Incurable Home The marriage of Robert Ludgate and Susanna Willis is listed in the NAI Index of Marriage Licences for which no supporting documents have survived. Her death certificate shows her birth as abt. 1814. There is no birth of a Willis in Kilshannig around this time. William Willis married Anne Berry in Kilshannig in 1825. There is no other Willis in Kilshannig until 1832 when an Elizabeth Willis daughter of William Willis and Catherine was baptised. William Willis was "received" into the church in 1834 and was described as "late of the 22nd Regiment". William and Catherine had 7 children in Scarragh including a daughter baptised Susan in 1839. The 22nd foot were in Southwest Ireland from 1822 until around 1830, which included patrols to Mallow. There were, however two William Willis in the 22nd at that time, one born Swindon and one born Enniscorthy. The William from Swindon enlisted around 1812 and the William from Enniscorthy in 1825.
William and Catherine had 2 earlier children in Tralee Barracks, William (b. 1829) and Francis (b 1830). Francis seems to have been baptised in both the Church of Ireland and Roman Catholic Church. A second William was born 1829 to William and Anastasia Willis. William and Anastasia Quinnwere married in Abbeyfeale, Co Limerick.
There are a number of Willis families in Co Cork including Rathcormack about 3 miles from Kilshannig.

Robert Ludgate and Susanna Willis had the following children:

     i.   **Thomas Ludgate** was christened on 29 Jul 1838 in Kilshannig by Mallow, Cork, Ireland.[67] He was buried on 17 Oct 1838 at Scarragh in Kilshannig by Mallow, Cork, Ireland.[68]

2   ii. **Michael Edward Ludgate**, died 26 Jan 1923, Belfast, Antrim, Northern Ireland.

   iii. **Mary Alice "Alice" Ludgate** was christened on 26 Jan 1842 in Kilshannig by Mallow, Cork, Ireland.[69] She died on 28 May 1865 at Cork Lunatic Asylum in Cork, Ireland.[70]

   iv. **Elizabeth Ludgate** was christened on 28 Sep 1843 at Scarragh in Kilshannig by Mallow, Cork, Ireland.[71] She was buried on 17 Nov 1856 at Scarragh in Kilshannig by Mallow, Cork, Ireland.[72]

   v. **Robert Ludgate**[73] was christened on 13 Oct 1845 in Kilshannig by Mallow, Cork, Ireland.[74] He was buried on 25 Nov 1856 at Scarragh in Kilshannig by Mallow, Cork, Ireland.[75]

   vi. **Richard Ludgate** was christened on 19 May 1847 in Kilshannig by Mallow, Cork, Ireland.[76] He died Cerebral Disease on 18 Jun 1880 at Cork District Lunatic Asylum in Cork, Cork, Ireland.[77] Batchelor

   vii. **David Ludgate**[78] was christened on 6 Aug 1849 in Kilshannig by Mallow, Cork, Ireland.

   viii. **William Joseph Ludgate** was christened on 7 Jul 1853 in Kilshannig by Mallow, Cork, Ireland.[79] In

---

63. , Mallow Heritage Centre/Familysearch.
64. , Griffiths Valuation Cork Sheet No 32 Map Ref 4.
65. , (: ,), FMP Ireland Diocesan And Prerogative Marriage Licence Bonds Indexes 1623-1866.
66. , , , .
67. , Familysearch Film FHL 874437 Item 7.
68. , Mallow Church of Ireland Registers.
69. , Familysearch Film FHL 874437 Item 7.
70. , Cork Registration District 1865 vol 2/10/82.
71. , Mallow Heritage Centre Index/Familysearch.
72. , Mallow Church of Ireland Parish.
73. , Familysearch Film FHL 874437 Item 7.
74. Ibid.
75. , Mallow Church of Ireland Parish Registers.
76. , Familysearch Film FHL 874437 Item 7.
77. , , Cork Registry 1888 page 94 Line 172, .
78. , Familysearch Film FHL 874437 Item 7.
79. Ibid.

From: *Percy E. Ludgate Prize in Computer Science*, TCD-SCSS-X.20121208.002, 11th October, 2019. © 2019, The John Gabriel Byrne Computer Science Collection

Extracts of Percy Ludgate's Genealogy

1887–1932 he was a Journalist in Cork, Cork, Ireland.[80] He died on 25 Nov 1936 at Victoria Hospital in Cork, Cork, Ireland.[81] Widower age 84, journalist, Senile Decay, myocardial congestion. Informant Gertrude Taylor, Occupier Victoria Hospital

6. **Thomas McMahon** was born about 1803 in Kilmore, Armagh, Ireland. He served in the military Attested to the Royal Artillery on 26 May 1825 in Armagh, Armagh,.[82] He served in the military Royal Sappers & Miners in 1841 in Woolwich, Kent, England.[83] WO11 /92 Hythe 7th Co Engineers and Miners 01-Jan 31-Mar 1841 Muster Thomas McMahon Miner
WO11 /92 Hythe 7th Co Engineers and Miners 01-Apr 30-Apr 1841 Muster Thomas McMahon Miner
WO11 /92 Hythe 7th Co Engineers and Miners 01-May 31-May 1841 Muster Thomas McMahon Miner
WO11 /92 Hythe 7th Co Engineers and Miners 01-May 31-May 1841 Muster Thomas McMahon Miner
WO11 /92 Woolwich 7th Co Engineers and Miners 01-Jun 30-Jun 1841 Muster Thomas McMahon Miner
WO11 /92 Woolwich 7th Co Engineers and Miners 01-Jun 30-Jun 1841 Thomas McMahom transfer to RHA
WO11 /92 Woolwich 7th Co Engineers and Miners 01-Jul 01-Jul 1841 Thomas McMahom transfer to Regiment of Artillery
WO11 /92 Woolwich 7th Co Engineers and Miners 01-Jul 01-Jul 1841 Thomas McMahom transfer to Regiment of Artillery
Thomas appeared in the census in 1841 at East Street in Plumstead, Kent, England.[84] Occupation Soldier
Age 35
Birth year 1806
He served in the military Admitted to the Pension List on 10 Oct 1843 in Woolwich, Kent, England.[85] Age 40
Total Service 18 years 6 months
Rate of Pension 9d
Cause of Discharge Chronic cough & palpitations of the heart
Place of Birth Kilmore, County Armagh
Place of Residence London He served in the military Discharge on 10 Oct 1843 in Woolwich, Kent, England.[86] Born in the Parish of Killmore in the county of Armagh. Attested age 22, trade a weaver Thomas died before 1849 at the age of 46. Probably died in the Royal Hospital, Chelsea Thomas McMahon's year and place of birth is derived from his Military Records. Church of Ireland records for Killmore start 1789 but no McMahon. Earliest records for other denominations start 1815. Thomas McMahon is a common name throughout Ireland. Frances "Fanny" Reed and Thomas McMahon were married on 4 Nov 1834 at Saint Luke's in Charlton, Greenwich, Kent, England.[87]

7. **Frances "Fanny" Reed** was born about 1811 in Ireland. Possibly Kinsale or Ballincollig where he father may have been stationed, Living at this address[88] She appeared in the census in 1851 at Royal Hospital Chelsea in Chelsea, Middlesex, England.[89] Marital status Married
Gender Female
Age 40
Birth year 1811
Occupation Nurse
Birth place Ireland
She appeared in the census in 1861 at Royal Hospital Chelsea in Chelsea, Middlesex, England.[90] Nurse age 50 Fanny appeared in the census in 1871 at Bowater Crescent in Woolwich, Kent, England.[91] Annuitant Age 60 born Ireland living with:
Frances M Clarke Daughter - Female 17 1854

---

80. , , , .
81. Ibid.
82. , National Archives UK WO 69/8.
83. , , National Archives WO11/92.
84. Census, , .
85. , , FMP National Archives WO23/15.
86. , National Archives UK WO 69/8.
87. , FMP Thames & Medway Marriages.
88. Census, , .
89. Ibid.
90. Ibid.
91. Ibid.

She appeared in the census in 1881 at Upper Market Street in Woolwich, Kent, England.[92] Living with William Scott and Wife a Pensioner
Relationship Lodger Head
Marital status Widow
Gender Female
Age 71
Birth year 1810
Occupation Pensioner
Birth town -
Birth county -
Birth place Ireland
Living with daughter, Augusta, single a domestic servany

Thomas McMahon and Frances Reed had the following children:

    i. **Augusta McMahon** was born on 13 May 1834 in Woolwich, Kent, England.[93] She was christened on 7 Sep 1836 at Saint Mary Magdalene in Woolwich, Kent, England.[94] She appeared in the census in 1841 at East Street in Plumstead, Kent, England.[95] Augusta appeared in the census in 1851 at Belgrave Road, Saint Georges Hanover Square, in London, England.[96] First name(s) Augusta M
Last name Mabone
Relationship Servant
Marital status Unmarried
Gender Female
Age 16
Birth year 1835
Occupation House Servant

    ii. **Thomas McMahon** was born on 7 Apr 1836 in Woolwich, Kent, England.[97] He was christened on 7 Sep 1836 at Saint Mary Magdalene in Woolwich, Kent, England.[98] He appeared in the census in 1841 at East Street in Plumstead, Kent, England.[99]

    iii. **Arthur McMahon** was born on 28 May 1838 in Woolwich, Kent, England.[100] He was christened on 24 Jun 1838 at Saint Mary Magdalene in Woolwich, Kent, England.[101] He appeared in the census in 1841 at East Street in Plumstead, Kent, England.[102]

3   iv. **Mary Ann McMahon**, born 19 Nov 1840, Iden, Sussex, England, United Kingdom; died 22 Aug 1936, Dublin, Ireland.

    v. **Robert McMahon** was born on 18 Aug 1843 in Woolwich, Kent, England.[103] He was christened on 17 Sep 1843 at Saint Mary Magdalene in Woolwich, Kent, England.[104]

---

92. Ibid.
92. Ibid.
93. , Included on Baptism Record.
94. , Familysearch Indexing Project C05594-2.
95. Census, , .
96. Ibid.
97. , Entry in Parish Register.
98. , Familysearch Indexing Project C05594-2.
99. Census, , .
100. , Entered into Register.
101. , Familysearch Indexing Project C05594-2.
102. Census, , .
103. , Entered in Baptism register.
104. , Family Serach.

From: *Percy E. Ludgate Prize in Computer Science*, TCD-SCSS-X.20121208.002, 11th October, 2019. © 2019, The John Gabriel Byrne Computer Science Collection

Extracts of Percy Ludgate's Genealogy

Fourth Generation

8. **John Ludgate** was christened on 29 Feb 1752 at Scarragh in Kilshannig by Mallow, Cork, Ireland.[105] He was buried on 6 Jan 1837 at Scarragh in Kilshannig by Mallow, Cork, Ireland.[106] Elizabeth Farmar and John Ludgate were married in 1778 in Cork, Ireland.[107]

9. **Elizabeth Farmar** was christened on 23 Apr 1758 at Scarragh in Kilshannig by Mallow, Cork, Ireland.[108]

John Ludgate and Elizabeth Farmar had the following children:

      i.   **Mary Ludgate** was christened on 25 Jul 1779 at Scarragh in Kilshannig by Mallow, Cork, Ireland.[109]

      ii.   **Katherine Ludgate** was christened on 25 Nov 1781 at Scarragh in Kilshannig by Mallow, Cork, Ireland.[110]

      iii.   **Matthew Ludgate** was christened on 10 Jun 1784 in Kilshannig by Mallow, Cork, Ireland.[111]

      iv.   **Michael Ludgate** was christened on 12 Jul 1786 at Scarragh in Kilshannig by Mallow, Cork, Ireland.[112]

      v.   **Elizabeth Ludgate** was christened on 17 Oct 1790 at Scarragh in Kilshannig by Mallow, Cork, Ireland.[113]

      vi.   **John Ludgate** was christened on 24 Feb 1793 at Scarragh in Kilshannig by Mallow, Cork, Ireland. He was buried on 6 Sep 1794 in Kilshannig by Mallow, Cork, Ireland.[114]

      vii.   **Jane Ludgate** was christened on 22 Mar 1795 at Scarragh in Kilshannig by Mallow, Cork, Ireland.[115]

      viii.   **David Ludgate** was christened on 2 Apr 1797 at Scarragh in Kilshannig by Mallow, Cork, Ireland.[116] Married Jane Berry 1824 (FamilySearch Indexing Project (Batch) Number M70035-1) Parents of David Ludgate (b 1838) and Robert Ludgate (b 1836). who joined the North Cork Militia/21st Foot with Michael Edward Ludgate. David joined the Civil Service in 1877. (British Civil Service Evidence Of Age held by the Society of Genealogists)

      ix.   **John Ludgate** was christened on 19 May 1799 at Scarragh in Kilshannig by Mallow, Cork, Ireland.[117]

      x.   **Margaret Ludgate** was christened on 1 Jul 1801 at Scarragh in Kilshannig by Mallow, Cork, Ireland.[118]

4   xi.   **Robert Ludgate**.

      xii.   **Thomas Ludgate** was christened on 30 Mar 1804 at Scarragh in Kilshannig by Mallow, Cork, Ireland.[119]

---

105. , Mallow Heritage Centre.
106. , RootsIreland/Mallow Heritage Centre.
107. , FMP Ireland Diocesan And Prerogative Marriage Licence Bonds Indexes 1623-1866.
108. , Mallow Heritage Centre/Mallow CofI Parish.
109. , MallowHeritage Centre/Family Search.
110. , Mallow Heritage Centre/Familysearch.
111. Ibid.
112. Ibid.
113. Ibid.
114. Ibid.
115. Ibid.
116. Ibid.
117. Ibid.
118. Ibid.
119. Ibid.

4. **Felix Reed** was born about 1758 in Drummaul, Antrim, Ireland.[120] Parish registers only exist from 1823 He served in the military Enlist 8th Battalion Royal Artillery on 16 Mar 1777.[121] He served in the military Promotion to Master Gunner 7th Battalion on 22 Feb 1805 in Ballincollig, Cork, Ireland.[122] Felix served in the military To Pension on 16 Jan 1812 at Fort Charles in Kinsale, Cork, Ireland.[123] He died on 16 Jan 1833 at the age of 75 at Ballincollig Military Cemetery in Ballincollig, Cork, Ireland.[124] Died when 77 years old. Served as Master gunner.

He was also known as Reid. Service Records WO97 /1257 at the UK National Archives gives Felix Reed/Reid date of birth as about 1758 in Drummaul, County Antrim. The earliest Parish Registers for Drummaul (Church of Ireland) are 1823.

There were Reed/Reids living in this area at the time of the Griffiths Valuation. Ann Cropley and Felix Reed were married on 9 Jul 1793 at St Alphege in Greenwich, Kent, England.[125]

15. **Ann Cropley** was christened on 20 Oct 1771 at St. Mary's in Chatham, Kent, England.[126]

Felix Reed and Ann Cropley had the following children:

- i. **Mary Ann Reed** was christened on 6 Oct 1799 in Norwich, Norfolk, England.[127]
- ii. **Felix Reed** was christened on 5 Aug 1801 at Saint Thomas in Portsmouth, Hampshire, England.[128] Possibly private in the 67th Foot. Died Poona, Bombay 29 Jul 1824
- 7   iii. **Frances "Fanny" Reed**, born abt 1811, Ireland.

120. , Service Records WO97 /1257 National Archives.
120. , Service Records WO97 /1257 National Archives.
121. , Service Records WO69/824 National Archives.
122. , Service Records WO 69/618 National Archives.
123. , Service Records WO97 /1257 National Archives.
124. , , Find A Grave Memorial# 143414831.
125. , London Metropolitan Archives P78/ALF item 034.
126. , Familysearch FHL microfilm 1469178..
127. , Familysearch Film 1526854.
128. , Familysearch Film 919726.

Fifth Generation

16. **Matthew Ludgate** appeared in the census in Apr 1766 at 1766 Religious Census in Kilshannig by Mallow, Cork, Ireland.[129] **1766 RELIGIOUS CENSUS OF KILSHANNIG**
Protestant Families - 61. Popish Families - 409. Total - 470
One Popish Priest - David Cahil - and no Fryer that I can hear of. - Given under my hand this 21, April, 1766 - James Hingston, Curate
Pr. – Protestant
Ludgate, Mattw. - **Pr.** He was buried on 30 Jan 1778 at Scarragh in Kilshannig by Mallow, Cork, Ireland.[130] Baptismal records for Kilshannig start in 1731 with both Robert and Elizabeth probably being born before this date.

17. **Mary** was buried on 25 Jan 1777 at Scarragh in Kilshannig by Mallow, Cork, Ireland.[131]

Matthew Ludgate and Mary had the following children:

      i.   **Elizabeth Ludgate**[132] was christened on 20 Jul 1735 in Kilshannig by Mallow, Cork, Ireland.
      ii.   **David Ludgate**[133] was christened on 31 Jan 1736 in Kilshannig by Mallow, Cork, Ireland.
      iii.   **Jane Ludgate** was christened on 4 Feb 1740 at Scarragh in Kilshannig by Mallow, Cork, Ireland.[134]
      iv.   **Ann Ludgate** was christened on 11 Mar 1743 at Scarragh in Kilshannig by Mallow, Cork, Ireland.[135]
      v.   **Mary Ludgate** was christened on 30 Oct 1746 in Kilshannig by Mallow, Cork, Ireland.[136]
      vi.   **Mary Ludgate** was christened on 4 Oct 1747 in Kilshannig by Mallow, Cork, Ireland.[137] Burial Page 247 of Register
      vii.   **Margaret Ludgate** was christened on 3 Dec 1749 at Scarragh in Kilshannig by Mallow, Cork, Ireland.[138]
8   viii.   **John Ludgate**.
      ix.   **Robert Ludgate** was christened on 31 Aug 1755 at Scarragh in Kilshannig by Mallow, Cork, Ireland.[139]

18. Eldest child, George was baptised in 1745. A possible earlier child, John, was baptised in 1742. Baptismal records for Kilshannig start in 1731 with both Robert and Elizabeth probably being born before this date. No marriage date has been found to establish Elizabeth's family name.
There were a number of Farmer families in Kilshannig and County Cork at this time.
Robert Farmar was born (date unknown). Elisabeth Boulster and Robert Farmar were married on 20 Aug 1741 in Kilshannig by Mallow, Cork, Ireland.[140] Connel formerley Boulster Widow

19. **Elisabeth Boulster** was buried on 5 Aug 1762 at Scarragh in Kilshannig by Mallow, Cork, Ireland.[141]

Robert Farmar and Elisabeth Boulster had the following children:

---

129. , Religious Census 1766 Return No 1123 Reference M 5036 (a).
130. , RootsIreland/Mallow Heritage Centre.
131. Ibid.
132. , Familysearch Film 596421.
133. Ibid.
134. , Mallow Heritage Centre/Familysearch.
135. Ibid.
136. Ibid.
137. Ibid.
138. Ibid.
139. Ibid.
140. , FMP Ffolliott Collection.
141. , RootsIreland/Mallow Heritage Centre.

---

i. **George Farmar** was christened on 9 Jun 1745 at Scarragh in Kilshannig by Mallow, Cork, Ireland.[142]

ii. **Ellen Farmar** was christened on 6 Mar 1747 at Scarragh in Kilshannig by Mallow, Cork, Ireland.[143]

iii. **Mary Farmar** was christened on 6 Jan 1750 at Scarragh in Kilshannig by Mallow, Cork, Ireland.[144]

iv. **John Farmar** was christened on 12 Aug 1753 at Scarragh in Kilshannig by Mallow, Cork, Ireland.[145]

9   v. **Elizabeth Farmar**.

vi. **Margaret Farmar** was christened on 27 Apr 1760 at Scarragh in Kilshannig by Mallow, Cork, Ireland.[146]

30. **Thomas Cropley** was born about 1739 in Tilbury Juxta Clare, Essex, England.[147] He served in the military Royal Navy Record in 1760–1827 in Greenwich, Kent, England.[148] ABT. 1760-1775

Calculated on Length of Service from ADM 73/54

ADM 73/206/12
1772
Muster as Quality Sergeant
HMS Dispatch

ADM 36/10887
1791
Muster as Corporal
HMS Courageux

ADM 73/5/353
1792
Admitted to Greenwich Hospital dates served: 3 March 1775-16 September 1790

ADM 73/54
1792
Rough Entry Book of Pensioners Living Woolwich 30 years' service profession Weaver Children 13, 9, 3 ,2
HMS Courageux

ADM 73/38
1795
Discharged Greenwich Hospital

ADM 73/54
1798
3 sons 2 daughters 40 Years' Service Sailor and Marine
HMS Repulse

ADM 27/6/409
1798
Allotment declarations Pay Book No 22 Master at Arms. Mary at Chatham
HMS Venerable

ADM 73/125
1798

---

142. Ibid.
143. Ibid.
144. Ibid.
145. Ibid.
146. Ibid.
147. , Service Records National Archives UK ADM 73/54 and others.
148. , National Archives UK ADM Series.

Out-pension Pay Book

ADM 27/6
1801
Allotment declarations Pay Book No 15 Mary at Chatham, 3 Sons 2 Daughters
HMS Standard

ADM 27/6/145
1801
HMS Standard; Pay book number: SB 15; Rank: Master at Arms Discharged invalided
HMS Standard

ADM 6/272
1801
Register of applicants to Greenwich Hospital for admission, out-pensions or other relief Out Pensioner

ADM 73/129
1801
Out-pension Pay Book Lady Day

ADM 73/65
1802
British Royal Navy & Royal Marines Service And Pension Records

ADM 73/55
1802
Rough Entry Book of Pensioner Last Residence Fair Row 40 years in Kings Service,
HMS Standard

ADM 73/38
1802
General Entry Book of Officers and Pensioners Died 10/Mar/1827

 He was buried on 16 Mar 1827 at St Alphege in Greenwich, Kent, England.[149] Resident at Greenwich Hospital
Admiralty records show Thomas Cropley's year of birth to be born abt. 1739 in Tilbury, Essex, with the exception of
ADM 73/43 which gives his date of birth as 1746 in Claire, Suffolk. At this time modern Tilbury did not exist and was
part of the parish Chadwell St Mary. The parish of Tilbury Juxta Clare in Essex is about 3 miles from Clare.
The registers for Tilbury Juxta Clare for this period have not survived.
The name Cropley can be found in Suffolk parish registers from the 16c as well as the neighbouring parts of Essex and
Cambridgeshire. Mary Lowrey and Thomas Cropley were married on 17 Oct 1770 at St. Mary's in Chatham, Kent,
England.[150]

31. **Mary Lowrey** was christened on 5 May 1749 at St. Mary's in Chatham, Kent, England.[151] She was buried on 19 Feb
1816 in Chatham, Kent, England.[152] Parents Alexander and Mary Lowry.
A Mary Lawry Bapt. 9th December in Chatham,the daughter of Thomas Lawry and Mary is a possibility.
Chatham is a busy naval port and, although the name Lowrey has existed in Kent for some time, the family may have
arrived in the area recently.

Thomas Cropley and Mary Lowrey had the following children:

>   15   i.   **Ann Cropley**.
>   ii.   **Alexander Cropley** was christened on 3 Oct 1773 in Chatham, Kent, England.[153]
>   iii.   **Thomas Cropley** was christened on 19 Jul 1778 at Saint Mary Magdalene in Woolwich, Kent,

---

149. , St. Alphege Church Greenwich 1825-1837 Burials.
150. , Familysearch FHL Film Number:1473651.
151. , Familysearch Image Catalogue.
152. , FMP St.Marys Chatham 1813-1837 burials, Thames & Medway Burials.
153. , Famil;ysearch FHL microfilm 1473646..

England.[154] He served in the military Admitte to Greenwich Hosptal School on 16 Jul 1790 at Greenwich Hospital School in Greenwich, Kent, England.[155] He served in the military Prisoner of War in 1800.[156] Previous ship HMS Repulse

    iv. **Mary Cropley** was born on 7 Apr 1783 in Woolwich, Kent, England. She was christened on 27 Apr 1783 at Saint Mary Magdalene in Woolwich, Kent, England.[157]

---

154. , Ancestry Parish Registers.
155. , National Archives UK ADM 29/1-96.
156. , National Archives UK ADM 103/500.
157. , London Metropolitan Archives P97/MRY/009.

From: *Percy E. Ludgate Prize in Computer Science*, TCD-SCSS-X.20121208.002, 11th October, 2019. © 2019, The John Gabriel Byrne Computer Science Collection

Extracts of Percy Ludgate's Genealogy

Sixth Generation

60. **Thomas Cropley** lived Settlement Certificate in Tilbury Juxta Clare, Essex, England on 7 Jan 1741.[158] Thomas Cropley and Mary were the subject of a Settlement Certificate issued by Tilbury Juxta Clare to Clare in 1741 Mary Ellingham and Thomas Cropley were married on 17 Oct 1739 at St Peter and St Paul in Clare, Suffolk, England.[159] By Licence

61. **Mary Ellingham** was born (date unknown).

Thomas Cropley and Mary Ellingham had the following child:

    30   i. **Thomas Cropley**, born abt 1739, Tilbury Juxta Clare, Essex, England.

---

158. , Suffolk Archives FL501/7/270.
159. , Society of Genealogy Suffolk Parish Register Transcripts.

Extracts of Percy Ludgate's Genealogy

From: *Percy E. Ludgate Prize in Computer Science*, TCD-SCSS-X.20121208.002, 11th October, 2019. © 2019, The John Gabriel Byrne Computer Science Collection

9 781911 566298